The Forty-eighth Alabama Infantry Regiment, C.S.A., 1862-65

BY

Joshua Glenn Price

Joshua Price

To order an autographed copy of this book, please contact the author at the following address:

fortyeighthalabama_author@yahoo.com

The Forty-eighth Alabama Infantry Regiment, C.S.A., 1862-65

Copyright © 2017 by Joshua Glenn Price

All rights reserved.

Printed in the United States of America. No part of this book may be used or reproduced in any manner whatsoever without written permission except in the case of brief quotations embodied in critical articles and reviews.

Fifth Estate, Blountsville, AL 35031

First Edition (2010)

Second Edition (2017)

Alabama Bicentennial Edition, 2017

Cover Designed by An Quigley

Printed on acid-free paper

Library of Congress Control No: 2017945536

ISBN: 9781936533954

Fifth Estate, 2017

Joshua Price

The Forty-eighth Alabama Infantry Regiment, C.S.A., 1862-65

Alabama Department of Archives and History

This flag is an Army of Northern Virginia, 6 th wool bunting issue. It was manufactured at the Richmond Depot during the winter of 1864-65. The flag was apparently surrendered at Appomattox and eventually forwarded to the U.S. War Department in Washington. There, it was assigned Capture Number 392 and described as the "Battle-flag of the Forty-eight Alabama, Fields Division, Longstreet's Corps." The flag was returned to the State of Alabama effective March 25, 1905.

Joshua Price

The Forty-eighth Alabama Infantry Regiment, C.S.A., 1862-65

This work is dedicated to the memories of my ancestors who fought in the Forty-eighth Alabama Infantry Regiment, Confederate States of America

Sergeant James Madison Parrish, Company C

Sergeant Thomas Jefferson Parrish, Company C

Private William A.T. Parrish, Company C

Private Isaac M. Parrish, Company C

Private James Scruggs, Company C

Private Green Lewis Rowan, Company G

Joshua Price

TABLE OF CONTENTS

INTRODUCTION	1
1 A CALL TO ARMS	11
2 CEDAR RUN TO SHARPSBURG	23
3 FREDERICKSBURG TO SUFFOLK	53
4 NORTH TO GETTYSBURG	63
5 CAMPAIGNING IN TENNESSE	77
6 GRANT VERSUS LEE	97
7 THE FINAL ROLL	111
8 CONCLUSION	121
APPENDIX A – SELECT LETTERS	129
APPENDIX B – SELECT BIOGRAPHIES	159
APPENDIX C – COMPANY ROSTERS	173
APPENDIX D – BURIAL SITES	279
APPENDIX E – RUBEN EWING OBITUARIES	305
APPENDIX F – SHEFFIELD MURDER TRIAL	309
APPENDIX G – APPOMATTOX ROSTER	311
APPENDIX H – COMPANY CAPTAINS	317
APPENDIX I – SELECT PHOTOGRAPHS OF SOLDIERS	319
BIBLIOGRAPHY	343

Introduction

It is hard to believe that seven years have passed since I completed, expanded, published, printed, and circulated the first edition of my work on the Forty-eighth Alabama - a Confederate infantry regiment that was, at the time, so obscure I could hardly fathom any person having any possible interest in learning about. This work has been out of regular publication and circulation since 2012, primarily for my lack of time to promote it, but is being revived thanks to Amy Rhudy at the Blount County Heritage Museum in Oneonta, Alabama. A chance encounter with this lady in May 2017 has breathed life, once again, into the story of these brave soldiers of so many years ago.

I have kept with the original research, text, and content in this version of the work. I have changed nothing more than basic grammar. This edition is not in anyway different in terms of content than the original – this is only to revive the story and present it once again to those interested in Alabama history, Civil War history, family history, and those interested in a good human interest story.

In the Fall of 2001 I enrolled as an undergraduate at Jacksonville State University in Jacksonville, Alabama with the intent of obtaining a Bachelor's Degree in History. The first course I signed up for was a graduate level course on Civil War history. The professor, Dr. Jennifer Gross, was beginning her first semester teaching since the completion of her doctorate program only months earlier. One of the assignments for the course was a ten-page paper on any original topic as long as it pertained to Civil War history. Many original ideas ran through my head, but none were *unique*. I decided I would write about my Civil War ancestors.

Up to that point in time I was never interested in genealogy. The thought of having to trace down ancestors through censuses, marriage records, deeds, bibles, letters, and God knows what else was completely appalling. Besides, the paper was due in less than three months! Time was definitely a problem. The solution, however, was a no-brainer. I consulted the PhD. of my family's genealogy – Mr. Earnest E. Hill of Rainbow City, Alabama. Earnest, in his familiar manner, told me anything and everything I wanted to know about exactly who my Civil War ancestors were. When I asked him about which regiment they fought in, he told me the Forty-eighth Alabama Infantry Regiment. He took me to the Gadsden Library and showed me the microfilm containing all of their military records. I was amazed! The following weeks I examined thoroughly all the military papers of my ancestors that Earnest provided me. I wrote my paper and got an "A" (of course). Most importantly, I wanted to know more about the Forty-eighth Alabama.

I searched and searched and found nothing about the regiment. No major works were written exclusively on the regiment. I discovered Penny and Laine's *Law's Alabama Brigade in the War*

Between the Union and Confederacy and quickly read through that book over the Christmas holiday break. But I wanted more.

In January 2002 I began my search for the soldiers themselves. I scoured every cemetery I thought one might be located – making note of every one I found. The first ones I found were of my ancestors – the Parrish brothers of Company C. Their graves, along with a few other soldiers from the regiment, were at the same cemetery in Marshall County, Alabama. I was hooked. I must have traveled a thousand miles around looking for cemeteries! As the list of burial sites grew my interest in the regiment grew as well. I wondered what the men in those graves looked like (in life, of course). Again, I consulted Earnest. Before long pictures of the soldiers came into my possession. But that was not enough. Soon, letters were arriving in my mailbox from family members. I soon filled two boxes with stuff that I did not know what to do with!

In 2006 I met a gentleman in Guntersville, Alabama named Pete Sparks. Dr. Sparks, from Grant, Alabama, also traced his ancestors to the Forty-eighth Alabama. Dr. Sparks, a historian and very knowledgeable about the regiment, immediately began touring me around the old Warrenton community and any site around Guntersville that linked to the regiment. He told me many stories about the regiment and its colonel – Jim Sheffield. I decided then I wanted to tell the story of the Forty-eighth Alabama Infantry Regiment.

During the summer of 2008 I revisited the history department at Jacksonville State University to discuss with faculty exactly what I should do with all the stuff now archived in my closet. I met with my undergraduate advisor, Dr. Paul Beezley, and told him I wanted to write a book on a Confederate regiment. Dr. Beezley immediately consulted Dr. Llew Cook, J.S.U.'s military historian, and they told me the best thing for me to do is to enter the graduate program and write a thesis using the material in my "archives".

In January 2009 I began writing a thesis on Confederate conscription using the Forty-eighth Alabama as the subject. Dr. Beezley, Dr. Cook, and Dr. Gross advised me and without their assistance, their knowledge, and their foresight this product would never have developed. In June 2010 my master's thesis was completed, defended, approved, signed, and ready to go. The thesis, *A History of the Forty-eighth Alabama Infantry Regiment, C.S.A., 1862-65*, contained nowhere near all the material and information from my "archives". I quickly expanded it to include all my "old" stuff – like letters, burial sites, photographs, biographies of each soldier, and anything else I could put my hands on.

Missing from this book are two particular pieces of information that are, in my opinion, *vital* to a complete regimental history. I have made every possible attempt to locate the number of men that fought in the regiment during the major battles of the war – particularly those examined in this book. Casualty returns are in abundance in the Official Records, but mean nothing when they cannot be converted into a statistic. Please note that since the number of engaged could not be obtained, I chose to omit the statistical portion of this work. The second piece of information that is not present in this work is an explanation of weapons used by the soldiers. I could locate no ordnance records that listed the types of weapons distributed among the soldiers. The Forty-eighth Alabama was issued, as mentioned in chapter two, weapons scavenged from the battlefields of the Seven Days when they arrived in Richmond in July 1862. I have not found any record stating they were issued weapons at any point during the war. To avoid a major fallacy, I have decided to also omit any explanation of type of weaponry distributed among the regiment and I do not have enough information to speculate. These are perhaps the only major pieces missing in this work.

It is impossible to list every person who contributed to this work. First, I have to thank all the people involved who contributed any minor detail to this work. They are spread, like their ancestors of the Forty-eighth Alabama after the war, from Georgia to Oklahoma. Little tid-bits made for great stories. Thank you all for saving and sending your information for me to preserve.

Appreciation is also due to Judge Bobby Junkins of Etowah County, Alabama. Judge Junkins is a descendent of Colonel James Sheffield and he opened his personal files concerning the colonel for me to explore. Judge Junkins made all letters, papers, pictures, and articles concerning Sheffield possible.

I would like to express my appreciation to the librarian involved in helping me research this regiment. Ms. Linda Cain, who is one of my favorite people at Jacksonville State University, is the former (now retired) third floor librarian at the Houston Cole Library. Ms. Cain has endured a few years of me and my continuous aggravation in pursuit of the most particular academic details. However, she never lost patience, never lost her smile, and never lost her wonderful personality as she *made sure* I got what I needed. Thank you so much Ms. Cain!

It would be impossible to complete a work such as this without the assistance of the ladies and gentlemen at the Northeast Alabama Genealogical Society. Their files, located at the Nichols Library in Gadsden, Alabama, contain the best family records of just about any family in northeast Alabama. Without their assistance and interest in my subject I never would have been able to locate as much material as I have located. I am confident that I could still find more in their files if only I made the time! Thanks alot guys!

I would also like to express my gratitude to Wayne and Gloria Gregg. With their assistance I was able to speak with the grandson of one of the soldiers of the Forty-eighth Alabama. Wayne, Gloria, and I interviewed William Taliaferro Scott, grandson of Private Calvin Scott, Company C, in December 2008 in his home in Hokes Bluff, Alabama. It was an awesome experience to be able to speak with someone who knew one of the soldiers and also to meet such a wonderful man. Mr. Scott died, of old age, soon after the interview. Thank you so much Wayne and Gloria!

Next, I would like to express my appreciation to the history department faculty at Jacksonville State University – most especially my thesis committee. Dr. Phillip Koerper gave me great advice in approaching what was important in the historical use of genealogical materials and how to obtain them. I would like to express special appreciation to the committee of this thesis. I would like to thank Dr. Llew Cook, director of my thesis, for his guidance and expertise in the field of military history. His advice greatly influenced my work and no doubt allowed it to expand. I would also like to thank Dr. Jennifer Gross for her expertise and counsel in the field of mid-nineteenth century social and political history that allowed this project to develop to include those two topics. I would like to express my immeasurable appreciation to Dr. Paul Beezley. Dr. Beezley greatly influenced the politics within this work and the examination of southern culture before and after the war. He also endured much aggravation during my graduate program. Without his never ending patience and counsel this project would never have flourished. Without the guidance of these professors this work could never have flourished. I am proud to have earned two advanced degrees from their department under their guidance. A million thank yous!

I would like to take the time to recognize everyone over the past seven years who supported this work and promoted it for everyone to enjoy. I could never name every person who promoted it by word of mouth, by email, and by social media. Thank you all so very much!

I would like to express my appreciation to my cousin and good friend the distinguished Mr. Earnest Edward Hill of Rainbow City, Alabama. Earnest is a *skilled* historian and without his invaluable assistance this work would *never* have been completed. I look forward to working on another project with him and wish him the best of luck with his own work.

This book is a culmination of many years of interest in this particular subject. My interest in the American Civil War dates back to when I was a kid in the 1980's. Knowing my interest in the conflict, my parents, Jimmy and Carolyn Price, kept me supplied with fresh standards on people and battles concerning the Civil War throughout my childhood - not to mention trips to the battlefields! My aunt, Sharlotte Head, also kept me supplied with books and helped me tremendously with the thesis version.

My granddaddy, the late Ralph Parmon "R.P." Parrish (February 19, 1919-August 28, 2001), also an avid reader, kept me "informed" and "educated" - as he enjoyed reminding me. As I have mentioned numerous times to numerous people, my love for history was fueled by all of these people. My granddaddy always told me

"Study history! You never know where you are going if you don't know where you've been!" Thank you for that.

Special appreciation goes to my wife Alysia for many years of patience, support, and encouragement. She has trekked as many miles as I have searching for information to hopefully resurrect this regiment from the depths of obscurity. She has sacrificed as much as any one person to see this project completed. To appreciate my work is to appreciate her work. I look forward to our next project together and those for many years to come.

Joshua Price
June 12, 2017

Joshua Price

1
A CALL TO ARMS

"Our property, our wives, our children, our honor, are all at stake."
- *James Lawrence Sheffield, December 1860*

This work is a historical sketch of the Forty-eighth Alabama Volunteer Infantry Regiment from its conception in April 1862 until its surrender at Appomattox Court House in April 1865. It begins by examining the soldiers' motivation for enlistment. These soldiers were not among the initial enlistees after the firing on Fort Sumter. Explaining why they chose to enlist in 1862 may shed light on the thousands of other soldiers who eventually volunteered to fight, but did not jump at the call in 1861. Modern arguments in southern culture are re-examined and used to clarify personal motivation for military service in the Confederate Army. The political culture of the citizens of the contributing counties and how that culture influenced hundreds of young men to serve in the Confederate Army is thoroughly examined. While the first part of this study focuses more on the social and cultural history of the men of the Forty-eighth, the bulk of this study is an evaluation of the military history of the regiment. Many accounts, including previously untapped primary sources, are vital in the examination of this regiment's history and its reputation as an effective combat unit.

Though there are minor variations in the political backgrounds of the twelve counties that contributed men to the Forty-eighth, an examination of Marshall County's political culture represents the norm among them. Marshall County's rich political culture shared many traits with those of the Jacksonians generations earlier. The yeoman farmers and lower class people of Marshall County did not trust the Whigs nor the Republicans. The Whigs never gained much support in Marshall County and, by 1856, hardly existed in the county. The other non-Democrats were "Know-Nothings," but joined the Democratic faction and supported Buchannan in 1856. The politics of secession, including the Compromise of 1850, split the interests of the dominant local Democrats in the late 1850's, uniting to vote collectively only in the presidential elections. State and Congressional elections reflected the split among voters between the State's Rights delegates and the Union Democrat delegates.[1]

Economic concerns dominated the political interests of the voters of Marshall County in the presidential election of 1860. The slave owners who lived along the Tennessee River feared for their slave property if a Republican became president. The lower class did not believe a Republican administration that wanted increased transportation, such as railroads, in the area to promote economic

---

[1] Larry Joe Smith, *Guntersville Remembered* (Albertville, AL.: Creative Printers, Inc., 1989): page 32; Ellis E. Moody, *Historical Events of Marshall County, Alabama Prior to the Civil War* (Boaz, Alabama, 1939).

advancement, could be afforded.[2]

After Abraham Lincoln's election to the presidency in November 1860, Alabama Governor Albert Moore called for a secession convention. Moore wanted each county in the state to send two delegates to the Montgomery convention – although some counties only elected one. Each county seat held elections in December 1860 and voters selected the convention's candidates. Most northern Alabama counties favored the cooperationists and the southern Alabama counties favored the straight-outs. Three men ran for the two positions representing Marshall County: James Lawrence Sheffield, Arthur Campbell Beard, and Samuel King Rayburn. Sheffield and Beard, cooperationists, had supported Stephen A. Douglas. Rayburn, a straight-out, unlike most candidates in the surrounding counties, had supported John C. Breckinridge for president. The voters elected Sheffield and Beard to the convention.[3]

The seven counties that contributed companies to the Forty-

---

[2] Ellis E. Moody, *Historical Events of Marshall County, Alabama Prior to the Civil War* (Boaz, Alabama, 1939): page 4.

[3] Smith, *Guntersville Remembered*, page 32. Like many "prominent" men of his time, Sam Rayburn believed that honor and loyalty were instrumental in the primary duties of all public servants. Rayburn, who was the first mayor of Guntersville, stayed behind and campaigned in the interests of his two colleagues. The three men developed a friendship that lasted the rest of their lives.

eighth Alabama Infantry Regiment sent eleven representatives to the Secession Convention in Montgomery, Alabama in January 1861: eight cooperationists and three straight-outs. Cooperationists in Calhoun County met in Oxford in November 1860 to select their candidates. However, the people of Calhoun County elected two straight-outs to the convention.[4] In his speeches at the Convention, Calhoun County representative G.C. Whatley blasted the "Black Republican" party that "does not recognize property in slaves." Whatley was one of the most vocal straight-outs in the Convention.[5] Sheffield, Marshall County representative, vocally disagreed with Whatley. Sheffield opposed secession until it was inevitable. "I have opposed secession as long as opposition was of any avail." Sheffield believed the Douglas Party was the only party that could save the bonds of the Union and prevent secession.[6]

Although ardent secessionists, such as Whatley, harbored a deep hatred for any person from the South who either supported the "Yankee" government or favored cooperating with it to avoid secession from the Union and newspapers, such as the Montgomery *Weekly Advertiser*, regularly insulted the "home-grown Yankees"

---

[4] *Jacksonville Republican*, 19 November 1860; Clarence Phillips Denman, *The Secession Movement in Alabama* (Montgomery: Alabama State Department of Archives and History, 1933): pages 161-166.

[5] William R. Smith, *The History and Debates of the Convention of Alabama, 1861* (Montgomery, Alabama: White, Pfister & Company, 1861): page 24.

[6] James Lawrence Sheffield letter to A.G. Henry of Marshall County, Alabama, January, 1861. See the letter in its entirety in the appendix of this thesis.

vocal in Alabama politics. Many historians have proved Confederate nationalism as being un-unified. For example, many citizens from Marshall County viewed the secession movement as illegal. They viewed Lincoln's election as constitutional and all arguments against it unjustifiable. Even some slaveholders in Alabama opposed secession before it happened, fearing it would lead to the absence of all government in the South. The slaveholders proclaimed their opposition of any war to keep their slaves, preferring to see the peculiar institution die rather than fight to preserve it.[7]

Despite the opposition to secession from some delegates, the ordinance of secession passed the convention quickly. Before he signed the ordinance, Sheffield sent a letter to the people of Marshall County asking them if he should sign it. "Our property, our wives, our children, our honor, are all at stake", he explained, "We cannot abandon the State in the hour of its coming trials, without incurring the reproach of all posterity."[8] Sheffield feared an invasion of the south by northern armies if the ordinance passed. He expressed his concern for the safety of the people of the south and also the future generations of southerners. Although Sheffield had opposed secession, after the convention voted in favor of it he dutifully

---

[7] Margaret M. Storey, "Civil War Unionists and the Political Culture of Loyalty in Alabama, 1860-1861." *Journal of Southern History 69*, no.1 (2003): pages 71-106.

[8] James Lawrence Sheffield letter to A.G. Henry of Marshall County, Alabama, January, 1861.

campaigned in favor of the ordinance. Despite his political convictions, Sheffield's loyalty lay with his native Alabama. "I feel bound to take the side of my native State in any contest which might grow out of it…having drawn my first breath upon her soil, she being my mother country, I intended to stand by her…her foes should be my foes and I would assist her sons in repelling every attempt to invade her soil either domestic or foreign foe." [9]

The secession convention gave Governor Moore the authority to create an army for the protection of the State of Alabama. Immediately after secession was announced, wealthy planters began donating money to raise regiments as many young men flocked to the drums of war.[10] Sheffield, Sam Rayburn, and *Marshall News* co-editor Thomas Eubanks volunteered for military service in April 1861 and served as officers in the Ninth Alabama Infantry Regiment. [11]

Nearly simultaneously with their enlistment, Union forces under Ulysses S. Grant entered northern Tennessee and won victories at Forts Henry and Donelson. By 7 April 1862 his force of over a hundred and ten thousand had defeated the Confederate army at the Battle of Shiloh and prepared to attack either northern

---

[9] William R. Smith, *The History and Debates of the Convention of Alabama, 1861,* pages 356-358.

[10] William Warren Rogers, et. al., *Alabama: The History of a Deep South State* (Tuscaloosa: The University of Alabama Press, 1994): pages 187-188.

[11] Smith, *Guntersville Remembered*, pages 32-33. The Ninth Alabama served in the western theater.

Mississippi or northern Alabama. The Confederate government faced a serious dilemma. If the thousands of service obligation contracts signed by volunteers for one year back in the spring of 1861 expired without renewal, the Confederacy's ability to wage an effective war against the Union would be drastically reduced. To prevent this issue, the Confederate government issued the Conscription Act of 1862 - to take effect April 16, 1862. The new terms of enlistment under conscription, three years to war, subjected all able-bodied men aged eighteen to thirty-five to mandatory military service. The Confederacy was in desperate need of soldiers. Many who joined the Confederate Army in April 1861, such as Sheffield and Eubanks, left their units when their previous contract expired to join or raise other companies. No records of Sheffield's motivation to form his own regiment exist. Perhaps he recognized the dominance of cooperationist and unionist sentiment in Northeast Alabama precluded raising a regiment from there in 1861, but when these other factors emerged in 1862, the possibility of raising a regiment became much greater.[12]

Most local newspapers urged those subjected to being drafted to volunteer rather than being forced into military service. With Grant sitting so close to Alabama, propaganda shifted from being

---

[12] Albert Burton Moore, *Conscription and Conflict in the Confederacy* (New York: Hillary House Publishers, Ltd., 1963): pages 12-17; see also Douglas Clare Purcell, "Military Conscription in Alabama During the Civil War". *Alabama Review 34,* no.2 (April 1981): pages 94-106.

politically motivated to a more solemn theme of defense. "Defend Calhoun Co. from the Yankee invaders!" and "Union Army could become Terrorists!" are examples of headlines that circulated in the spring of 1862.[13] One article reported "the 'Lincoln Congress' wished to colonize and subjugate the South. Owners of slaves will be hanged and the negroes will be given the jobs formerly occupied by white farmers." The possibility of Yankee soldiers entering their state terrified the people of northeast Alabama.[14] Editors understood the power of propaganda. Joseph Frank and George Reaves explored the use of propaganda in both the North and the South in their work *Seeing the Elephant*. They used many examples from newspapers from both the northern and southern states. In their studies they concluded that no one particular factor motivated men to enlist in the military. The men who enlisted in the military in 1861 and 1862 did so for various reasons, including honor, the protection of property and families, and after 1862, the fear of conscription.[15]

Many scholars have examined the reasons why men are willing to fight and sacrifice their lives in war. James McPherson argues that many personal reasons contributed to Civil War soldiers' instinct to fight – including honor, duty to country, adventure, and in many cases defense of their homeland. Even though he does include

---

[13] *Jacksonville Republican*, March-April 1858.

[14] *Jacksonville Republican*, 2 April 1856.

[15] Joseph Allan Frank and George A. Reaves, *"Seeing the Elephant": Raw Recruits at the Battle of Shiloh* (New York: Greenwood Press, 1989): pages 30-37.

personal reasons as being a major contributor to each individual's motivation, he concludes that each man, both north and south, felt a patriotic obligation to fight for his country be it the United States or the Confederate States.[16]

Sheffield's task of raising companies from the previously listed cooperationist counties in northeast Alabama would have been difficult without proper motivation. Without the support of the yeomen class, Sheffield and the planter class could not wage a war against the Union army. He understood that each man, bound by contract of personal honor and manhood, must provide for his own family and home. To assure that each man's family received financial assistance during the soldiers' absence, Sheffield offered a bounty of fifty dollars per *volunteer* – paid after the completion of initial training at Auburn in June 1862. Sheffield's ability to pay such large sums from his own pocket is testimony to his wealth and, perhaps, his interest in securing his own economic interests along the northern banks of the Tennessee River.[17]

On 7 April 1862 men from Marshall and Blount counties met at Warrenton, the seat of Marshall County located a few miles north of Guntersville, and formed five companies - A, B, C, D, and E.

---

[16] James M. McPherson, *For Cause and Comrades: Why Men Fought in the Civil War* (New York: Oxford University Press, 1997).

[17] *Jacksonville Republican*, 5 June 1862. Note that the bounty only applied to volunteers. Any man brought into service by conscription was not entitled to bounty.

Soon more men arrived at Warrenton from Blount County and Company F mustered on 10 April. Companies G, H, and I, from Cherokee County, mustered later in the month of April.[18] It is not coincidence that the companies of what became the Forty-eighth Alabama Volunteer Infantry Regiment formed during the first weeks of April 1862. Moses Lee, a Jacksonville, Alabama native and Confederate veteran, returned to Calhoun County in March 1862 and campaigned for volunteers to form a company from that county. Lee expressed that he hoped to complete the raising of his company before the conscription laws went into effect. Lee raised his company at Mt. Polk, northwest of Jacksonville, in May 1862.[19]

During elections for officers, Sheffield won unanimous support for colonel of the regiment. All companies elected officers as soon as each officially mustered. Company A, from the Brooksville area of Blount County, elected Andrew Jackson Alldredge as its captain. He resigned due to illness in July 1862, and his brother Jesse J. Alldredge replaced him. Company B, from southeastern Marshall County, elected Thomas J. Burgess as its captain. Company C, from the hills of southwestern Marshall County, elected W.S. Walker as its captain. Company D, from the Guntersville area, elected Samuel Cox as captain. Company E, also

---

[18] William C. Oates, *The War Between the Union and the Confederacy and its Lost Opportunities* (Dayton, Ohio: Morningside House, Incorporated, 1985): pages 792-795; same information can also be found at the beginning of the first roll of the Confederate microfilm regimental and personnel records.

[19] *Jacksonville Republican*, 1 April 1862.

from northern Marshall County, elected former secession convention candidate Samuel K. Rayburn as its captain. He served until he was elected Brigadier General of militia. Company F elected Reuben Ellis as its captain. Company G elected John S. Moragne as captain. Company H elected R.C. Golightly as captain. Company I elected John W. Wiggonton as captain, and he served until promoted to major in 1864. Company K, was commanded by Captain Moses Lee.[20]

These men volunteered for honor, their concept of manhood, and the need to defend their homeland. Over the following three years they traveled far from home and became involved in the greatest military struggle ever seen on the continent. Mustered in, this variety of men, many of them cooperationists before the war, prepared to board a train in Guntersville that would take them to war.

---

[20] Oates, *The War Between the Union and the Confederacy,* pages 792-795.

Joshua Price

2

CEDAR RUN TO SHARPSBURG

"Jackson's men fight like Devils!"
-Private Henry Young, 2nd Wisconsin Infantry

Sheffield and his friend Thomas Eubanks left the officer ranks of the Ninth Alabama Infantry Regiment before the expiration of their 1861 contracts to return home to Marshall County and raise a regiment. To outfit this new regiment of volunteers and to provide for them during the war, Sheffield invested approximately $57,000 of his own money. Sheffield's investment is testimony not only to his accumulated wealth, but also his resilient commitment to a cause he had once so vigorously opposed. Sheffield authorized Eubanks to utilize the *Marshall Eagle* to promote the raising of new companies consisting of local men.[21]

The newly created companies moved by rail from Warrenton to Jacksonville. At Jacksonville the regiment took the rail south through Talladega, Columbiana, and then to Selma.[22] The regiment

---

[21] Smith, *Guntersville Remembered*, pages 32-35; *Confederate Military History, Volume VIII* (Wilmington: Broadfoot Publishing Company, 1987): pages 797-798.

[22] J.T. Lloyd, "Lloyd's Map of the Southern States Showing all their Railroads, their Stations & Distances, also the Counties, Towns, Villages, Harbors, Rivers,

arrived in Selma, Alabama from Jacksonville in late April 1862 and marched to Montgomery. There the regiment was delayed two days while it waited for a boat to take it to Auburn for training. At the capitol it cooked two days rations and listened to a speech by Governor Albert Moore. A group of young ladies, eager to get a glimpse of the new soldiers, cheered the Calhoun County men as they boarded their boat and headed to Auburn. The regiment arrived "safe and sound" at Auburn the next day.[23]

The regiment entered formal military training at Auburn during the first week of May 1862. During this time the raw recruits were taught the basics of military service including marching, formations, and how to handle weapons. Elected regimental and company officers, along with drill instructors assigned to the regiments, worked hard to quickly instill both military doctrine and discipline in the recruits. Most officers of each company had prior military service and conducted training in their own discretion – under supervision of the regimental officers. Even though rank was always held in respect, there was little social class distinction in the Confederate regiments. Company officers, unlike regimental officers, had personal relations with the men that served under them. Most officers were democratically elected by the men and were often

---

and Forts", New York, 1862. See the web link in the bibliography to the image of this map at the Library of Congress.
[23] *Jacksonville Republican*, 5 June 1862. This report, written by E.F. Ellis, included a description of Moses Lee: "Capt. Lee makes his men obey him – they all like him, he stands square up to them, and sees that they all get their rights."

brothers or cousins or good friends with the men. If an officer was respected it was because he earned it by gaining respect from the men due to his performance in training or in combat. Its initial combat training in Auburn only lasted only a couple of weeks at the most.[24]

The newly formed Forty-eighth Alabama Infantry Regiment, commanded by Colonel James L. Sheffield, mustered into service 22 May 1862 and placed under General Thomas J. Jackson's command. It was immediately sent to Virginia to join the Confederate Army of Northern Virginia as it fought McClellan's invading army around Richmond. On 14 June the regiment boarded a train at Auburn, transferred trains in Atlanta, and headed up the east coast through the Carolinas toward Virginia.[25]

The Forty-eighth Alabama Infantry Regiment arrived in Richmond during the afternoon of 21 June 1862. As the men exited the train and formed in the nearby fields they could hear the roar of cannon a very short distance to the northeast. It was rumored among the ranks that the Forty-

---

[24] Brent Nosworthy, *The Bloody Crucible of Courage: Fighting Methods and Combat Experiences of the Civil War* (New York: Carroll and Graf Publishers, 2003): pages 148-158.

[25] Nelda A. Simpson and William E. Simpson, eds, *The correspondence of John M. Anderson, private, C.S.A.* Letters dated 14 June 1862 and also 22 June 1862. Anderson's Sixty-one letters were privately published and undated by the Simpsons and can be found in the Alabama Room on the tenth floor of Jacksonville State University's Houston Cole Library. Henceforth noted as "John M. Anderson."

eighth Alabama would soon get its first taste of combat.[26] During the afternoon of 22 June three wagon loads of captured and salvaged arms were brought to the Forty-eighth Alabama's camp near the Richmond fairgrounds. The men were proud of their new weapons.[27] Third Brigade, commanded by William B. Taliaferro, consisting of the Tenth Virginia Infantry, the Twenty-third Virginia Infantry, the Thirty-seventh Virginia Infantry, the Forty-seventh Alabama Infantry, and the Forty-eighth Alabama Infantry began its move to Culpeper, Virginia after the conclusion of the Seven Days Battles in late July 1862. The three veteran Virginia brigades moved by rail from Richmond to Hanover Junction, then to Louisa Court House. The Virginia regiments camped and drilled for a few days south of Gordonsville until they marched toward the enemy on 6 August.[28] The three Virginia regiments were likely moved out before the Alabamians were attached to them. The Alabama regiments did not travel from Richmond to Culpeper by rail. Instead, the raw volunteers performed their first march under Thomas Jackson's command.

---

[26] John M. Anderson to Elizabeth P. Anderson, 14 June 1862; see also *ibid.*, 22 June 1862.

[27] *Anderson Letters*, 22 June 1862; John O. Casler, *Four Years in the Stonewall Brigade* (Marietta, Georgia: Continental Book Company, 1951): pages 96-97.

[28] James Huffman, *Ups and Downs of a Confederate Soldier* (New York: William E. Rudge's Sons, 1940): pages 57-58. Huffman was a soldier in the Tenth Virginia Infantry.

The campaign between Robert E. Lee's Army of Northern Virginia and General John Pope's Army of the Potomac opened at the Battle of Cedar Run, near Culpeper, Virginia, on 9 August 1862. During this fight the untested Forty-eighth Alabama Infantry Regiment, attached to Colonel A.G. Taliaferro's Third Brigade, received its first taste of combat.

McClellan's army was still east of Richmond on the Peninsula when Pope approached the Confederate capitol from the west. Jackson's command, which was located along the Mechanicsville Turnpike west of Richmond, was ordered to march north and west to Gordonsville to prevent Pope from occupying that town. Jackson arrived there on 19 July, immediately placed skirmishers, and sent cavalry on reconnaissance. Pope sent cavalry to Gordonsville on 2 August and Jackson did the same, and a small skirmish ensued. The Federal cavalry vacated the area within minutes. On 7 August Jackson sent three of his divisions, including William Booth Taliaferro's, to attack Pope's forces near Culpeper Court House. On the morning of 8 August the Confederates encountered another detachment of Federal cavalry south of Culpeper, northwest of Slaughter Mountain. Both Jackson and Pope spent the remaining hours of 8 August marching their forces toward Slaughter Mountain. The forced march and the hot Virginia sun took a toll on the

Confederates, and many men fell out of line, suffering from heat exhaustion and some from heat strokes. On 9 August the fighting intensified as forces concentrated upon the open fields near Slaughter Mountain and Cedar Run.[29]

A.G. Taliaferro's Third Brigade, comprised of five regiments, moved northeast along the Culpeper Road and entered the battlefield, practically exhausted from the marching, at approximately 5:00 p.m. south of the intersection of Culpeper Road and Crittenden Lane. Third Brigade's line of battle stretched southward along the eastern side of the Crittenden Lane. Jubal Early's Brigade emulated the formation of Third Brigade and connected its left to Third Brigade's right.

At an *estimated* strength of over four-hundred fifty soldiers the Forty-eighth Alabama Infantry occupied the extreme right of Third Brigade's line of battle and took position behind two artillery batteries that dueled with Federal guns positioned across the three thousand feet wide cornfield along the Mitchell's Station Road. With the

---

[29] U.S. War Department. *The War of the Rebellion: A Compilation of the Official Records of the Union and Confederate Armies. OR, S1, V12, P2*, pages 181-182, T.J. Jackson. William B. Taliaferro became division commander before the battle and Colonel A.G. Taliaferro, formerly of the Twenty-third Virginia Infantry, took command of Third Brigade. Please note that "Taliaferro" is properly pronounced "Tol-i-ver"; Horace H. Cunningham, *Doctors in Gray* (Baton Rouge: Louisiana State University Press, 1993): page 173. Many of Jackson's troops, while under his command, suffered strokes and exhaustion from forced marches. One Hundred twenty-eight volumes total, Washington, D.C., 1880-1901. Henceforth noted as *OR* followed by *Series, Volume, Part*, and pages.

Confederate advance in check, General Christopher Augur positioned sharpshooters from the United States Eighth and Twelfth Battalions approximately one thousand five hundred feet from the Confederate line of battle. Augur moved his line of troops slowly, taking half an hour to form his division and prepare it for the assault on the Confederate line.

It is important to understand the positions of the regiments of Third Brigade during the early stages of Augur's attack. The Tenth Virginia Infantry had originally occupied the extreme left of Third Brigade's line; however, the regiment was moved to the northern sector of the battlefield to reinforce the struggling Confederate line there. After the Tenth Virginia moved out, the Forty-seventh Alabama Infantry anchored the extreme left of Third Brigade's battle line. Considered by historians as the "twin regiment" of the Forty-eighth Alabama, the Forty-seventh Alabama was also inexperienced in battle. To the right of the Forty-seventh Alabama were (from left to right from the rear) the Thirty-seventh Virginia Infantry, the Twenty-third Virginia Infantry, and the Forty-eighth Alabama Infantry covering the right flank of the brigade.

By 5:45 p.m., Augur's division marched toward the Crittenden Lane and exchanged deadly volleys with Third Brigade. Both armies were exposed in open field maneuvers without the privilege of any natural defenses. After taking many casualties in the exchange with Augur, Third Brigade suffered a terrible collapse. At approximately 6:15 p.m., the Forty-seventh Alabama was under severe fire on its front (eastern) and left (northern) flank. Federal troops poured volley after volley of terrible musket fire into the raw Alabama troops, and the regiment retreated. Meanwhile, the right flank of Third Brigade, protected by Jubal Early's brigade, engaged in a bitter slugfest with Augur's division.[30] During this engagement Colonel James Sheffield and Major Enoch Alldredge of the Forty-eighth Alabama suffered severe wounds and both were taken from the field.[31]

Generations of historians have condemned the Forty-eighth Alabama Infantry in the events that led to the collapse of Third Brigade at Cedar Run. It is clear that the Forty-seventh Alabama retreated, but it is *not clear* that the Forty-eighth Alabama was responsible for the collapse of the

---

[30] Robert K. Krick. *Stonewall Jackson at Cedar Mountain* (Chapel Hill: The University of North Carolina Press, 1990), pages 175-201; for a short autobiography of Colonel Alexander G. Taliaferro see *Confederate Veteran*, Vol. XXIX (1921), pages 126-129.

[31] J. Gary Laine and Morris M. Penny, *Law's Alabama Brigade in the War Between the Union and the Confederacy* (Shippensburg: White Mane Publishing Company, 1996), pages 31-37. This book examines the history of all five regiments of General Law's Brigade – the Fourth Alabama, the Fifteenth Alabama, the Forty-fourth Alabama, the Forty-seventh Alabama, and the Forty-eighth Alabama infantry regiments.

remainder of Third Brigade's battle line. The main source of the accusation used by historians is the report filed by Major H.C. Wood of the Thirty-seventh Virginia Infantry. Major Wood reported that the left of Third Brigade's battle line was protected by the Forty-seventh Alabama and the Forty-eighth Alabama regiments. Major Wood concluded, "...this regiment [the Thirty-seventh Virginia] would have been able to maintain its position had the Forty-seventh and Forty-eighth Alabama regiments had been able to maintain theirs." In his report, Major Wood did not acknowledge the Forty-eighth Alabama was positioned at the right of Third Brigade. Wood was the only reporting officer of Third Brigade to place the Forty-eighth Alabama on the left of Third Brigade's battle line with the Forty-seventh Alabama. He did not indicate which regiment was to the right of the Thirty-seventh Virginia.[32]

A member of the Tenth Virginia Infantry, who had been deployed north of Crittenden Lane, recalled "...two regiments of North Carolina conscripts had been placed in our Brigade. They had not been drilled and ran like turkeys when they got into the heat of this battle...Raw men can't stand that kind of music." The two regiments he referred to

---

[32]   *OR, S1, V12, P2*, page 206. Major Wood's report, most notably the quote used above from the report, is acknowledged in most books and articles of the Battle of Cedar Run. Other reports should be examined more closely.

were not conscripts from North Carolina, but rather the two regiments of Alabama volunteers. Unless this soldier was a straggler, he was with his regiment north of the Crittenden Lane fighting in that sector of the battlefield and would have had only second-hand knowledge of the engagement of the regiments of his brigade at Cedar Run. This soldier even admitted that "...I am now writing mostly from memory and my memory fails me..." Indeed, his reflections are likely distorted.[33]

Taliaferro's report reflected that of Major Wood. The colonel blamed the two Alabama regiments for the collapse after having given up their positions. But, Taliaferro did not take into consideration the reports of any officer other than Major Wood. Taliaferro had been in command of the Twenty-third Virginia Infantry until General Charles Winder was mortally wounded during early stages of the battle. Taliaferro took command of Third Brigade for most of the battle and was responsible for the formation of Third Brigade's line of battle along the Crittenden Lane. He did not include in his report to division headquarters the order of the regiments along Third Brigade's battle line.[34]

The Forty-seventh Alabama Infantry, guarding the

---

[33] Huffman, *Ups and Downs of a Confederate Soldier*, pages 58-59.

[34] *OR, S1, V12*, pages 206-207, Taliaferro.

left of Third Brigade's line of battle, came under intense fire on both their front and their exposed left. Colonel James W. Jackson, commander of the Forty-seventh Alabama, reported that the regiment fell into confusion when it received a murderous crossfire after being flanked. This retreat had a domino effect upon the line, starting from Third Brigade's left. All commanders placed the Forty-seventh Alabama at the left end of the line, but none can agree, including the Twenty-third Virginia's commander, that the Forty-eighth Alabama was connected to the Forty-seventh Alabama.[35]

Lieutenant Colonel Simon T. Walton, commander of the Twenty-third Virginia Infantry after Colonel Taliaferro's battlefield promotion, reported twice that the Thirty-seventh Virginia Infantry was to his regiment's left. He was ordered to retreat to prevent crossfire on his front and his left. He reported that his regiment was ordered to retreat after the Thirty-seventh Virginia Infantry fell out of line. If his reports were accurate the Forty-eighth Alabama Infantry would have been on his right covering Third Brigade's flank. The Forty-eighth Alabama would thus be cleared of any accusations for the collapse of the left of Third Brigade.[36]

Lieutenant Colonel Abner A. Hughes took command

---

[35] *OR, S1, V12*, pages 207-209, Jackson.

[36] *OR, S1, V12*, page 211, Walton.

of the Forty-eighth Alabama Infantry early in the battle after Colonel Sheffield was "painfully" wounded. After the battle, he reported that the regiment was ordered to the brow of a hill where the enemy lay exposed below. The Forty-eighth fired into them "without cessation". The Forty-eighth Alabama was located to the right of the Twenty-third Virginia Infantry and it was forced to fall back when the regiment began receiving concentrated fire on its left. He reported that his men "behaved gallantly", and his report gave no indication that his right, protected by Jubal Early's brigade, was ever in danger.[37] Colonel Sheffield did not submit an official report of the battle of Cedar Run, but he did write a report to the editor of the *Jacksonville Republican* describing the actions of Company K. Sheffield reported that Company K suffered more severely than any of the other companies at Cedar Run and took the heaviest casualties because Captain Moses Lee did not hear the call for retreat.[38] "We was verry well inisuated" recalled Captain Alvin Oscar Dickson of Company A.[39]

---

[37] *OR, S1, V12,* page 209, Hughes.

[38] *Jacksonville Republican*, 28 August 1862.

[39] Alvin Oscar Dickson to Montgomery Archives. Handwritten letter dated 8 October 1913. Captain Dickson served in Company A. from Blount County, Alabama. Dickson's letter to Owen, the latter being Director of Archives in Montgomery, Alabama, was written in response to Owen's request for information about the services of the Forty-eighth Alabama in the Confederate Army.

By 6:30 p.m., all of Third Brigade had retreated five hundred feet to the west beyond the Crittenden Lane and the Federal Army was in quick pursuit. Yet, Third Brigade rallied and immediately returned to the fight. By 6:45 p.m., Third Brigade had re-established its former position east of the Crittenden Lane, and the Federal advance was repulsed.

The Confederate lines north of the Culpeper Road managed to hold, and the reformation of the Confederate lines south of the Culpeper Road repulsed the last major Federal infantry threat. But the Federals were not finished. Against the advice of most of his subordinates, Pope ordered a cavalry assault across the cornfield south of the Culpeper Road toward the newly reformed battle lines of Third Brigade along the Crittenden Lane. The First Pennsylvania Cavalry made the charge. The Pennsylvania horsemen made easy targets for the regiments of Third Brigade. Lieutenant Colonel Hughes reported that the Forty-eighth Alabama Infantry, along with the rest of Third Brigade, drove the enemy "at every point" until darkness. The regiment slept on the battlefield that night.[40]

Dickson's original document is located at the Alabama Department of Archives and History in Montgomery, Alabama.

In the aftermath of the battle, the Forty-eighth Alabama Infantry recorded twelve men killed and sixty-one wounded, for a total of seventy-three casualties.[41] Major Enoch Alldredge was severely wounded in the calf in the early stages of the battle. Partially crippled from his wound, he resigned his commission and returned home in late October 1862. Colonel James Sheffield, also wounded severely in the leg, did not return to the regiment until September 18, 1862 – one day after the Battle of Sharpsburg.[42] Sheffield was not the only influential officer counted among the casualties on 9 August. Enoch Alldredge, the influential fifty-six year old major of Company A from Blount County, Alabama, was wounded in the right thigh – a wound that rendered him unable to serve and partially crippled for the remainder of his life. Alldredge's son Jesse, who achieved the rank of lieutenant colonel before the war ended, was wounded through both thighs at Cedar Run and also missed a considerable amount of time afterward in

---

[40] Krick, *Stonewall Jackson at Cedar Mountain*, pages 187-192; *OR, S1, V12, P2*, page 209, Hughes; see also, *ibid.*, pages 140-142, Falls' (First Pennsylvania Cavalry) report.

[41] *OR, S1, V12, P2*, page 179.

[42] Note that James Sheffield's entry in *Confederate Military History,* Volume VIII (Alabama), pages 797-798, has many errors. Sheffield did not serve as commander of Law's Brigade at the Battle of Sharpsburg. He was not active at the Battle of Sharpsburg because of the injuries sustained at Cedar Run (which was a leg wound, not a shell concussion – see his report in the *Jacksonville Republican*). Sheffield's regiment was not assigned to Law's Brigade until Jan. 1863.

recovery.[43]

The Battle of Cedar Run was a spectacular baptism of fire for the regiment. Sergeant John Dykes Taylor of Company D., Forty-eighth Alabama called the battle, "...one of the most terrific of the war for the time it lasted".[44] Private Calvin Scott of Company C, Forty-eighth Alabama, had three fingers shot off during the battle leaving only a thumb and a forefinger on his right hand. "He used to grab me by my overall straps for running through the house when I was a boy", recalled Scott's grandson, "it sure was hard to get away from his hook-hand!"[45] In June 1863, the Forty-

---

[43] Hoole, William Stanley, ed., *John Dykes Taylor's History of the 48th Alabama Volunteer Infantry Regiment, C.S.A.* (Tuscaloosa: Confederate Publishing Company, 1985): page 12. Henceforth noted as "*John Dykes Taylor.*" Originally published in The *Montgomery Advertiser* on 9 March 1902. Taylor enlisted as a private in Company E Forty-eighth Alabama Infantry at Warrenton near Guntersville on 7 April 1862. Due to his experience in the warehousing and dry goods business and as an attorney (and perhaps some political influence), he quickly rose to Regimental Ordnance Sergeant. He served at this position until late 1864 when he returned home from the war and never reported back. Before his death at the age of fifty-two in 1888 he compiled a series of letters that were altogether a narration of the regiment's experiences from its conception in Guntersville to the bloody campaign between Grant and Lee in the summer of 1864. This series of letters, first published by the State of Alabama in 1902, became the official history of the Forty-eighth Alabama. For more information on the life of John Dykes Taylor, see Larry Joe Smith's *Guntersville Remembered* (Albertville: Creative Printers, Inc.: 1989).

[44] Hoole, *John Dykes Taylor*, page 12. The details of Major Enoch Alldredge's wounds are presented in his letter of resignation dated 21 September 1862.

[45] Interview with William Taliaferro Scott of Hokes Bluff, Alabama in December 2008. "W.T.", now since passed away, remembered his grandfather Calvin Scott vividly as a "hard worker" and a "kind man". Calvin Scott died 14

eighth Alabama Infantry returned to the Cedar Run battlefield while en route to Gettysburg. Taylor recalled the scene, "...our brigade was marched to the old battlefield of Cedar Run, the scene of the first engagement in which the 47th and 48th Alabama regiments were engaged. I, together with most of those who remained of the two regiments, examined the old field and with feelings of sadness, saw some of the bones of our dead comrades who fell there, which were exposed by the rains. Sheffield sent a detail to cover any exposed bones of our dead."[46]

On 10 August 1862 Thomas Jackson's command remained on the battlefield at Cedar Run, south of Culpeper, Virginia, and prepared for a second day of fighting. Although Pope's Army of Virginia expected the arrival of considerable reinforcements under the command of Nathaniel Banks, Pope decided to retreat. He immediately began moving his army northeast toward the protection of the Rappahannock River. He hoped to regroup and receive reinforcements before Lee could coordinate an attack that would include the corps of both Jackson and James Longstreet, and also J.E.B. Stuart's cavalry.[47] Following the Federal retreat, Jackson

---

May 1925 and is buried in Old Clear Creek Primitive Baptist Church cemetery in Marshall County, Alabama.

[46] Hoole, *John Dykes Taylor*, page 12.

[47] John J. Hennessy, *Return to Bull Run: The Campaign and Battle of Second Manassas* (New York: Simon and Schuster, 1993): pages 38-59. Scholars have recognized Hennessy's book as the standard on this campaign, the most comprehensive work published on the campaign and battle of Second Manassas.

ordered each brigade to have details bury the dead and gather weapons and ammunition. [48] On the morning of 11 August Jackson received a flag of truce from Federal burial details. Jackson allowed them to bury their dead until 5:00 p.m., at which time the truce ended.[49] Although Jackson held the battlefield after the fight he ordered his command to march toward Gordonsville, Virginia on the evening of 14 August. There, he would have the protection of the bluffs that overlooked the Rapidan River and await orders from Lee to advance.[50]

Jackson, under the protection of Stuart's cavalry, broke camp at Gordonsville, Virginia on 15 August and began pursuit of Pope's army. As it marched north Third Brigade was without rations because the supply wagons had not caught up. Jackson's entire command crossed the Rapidan River at Somerville Ford on 20 August and camped near the town of Stevensburg, Virginia. On 21 August Colonel A.G. Taliaferro's Third Brigade led the march toward Beverly's Ford on the Rappahannock River. As it approached the ford it engaged a large number of Federal skirmishers. Third Brigade, with the assistance of Stuart's horse artillery batteries, dispersed the stubborn Federal defenders and

---

[48] Casler, *Four Years in the Stonewall Brigade*, pages 104-105.

[49] *OR, S1, V12, P2*, page 184, T.J. Jackson's report.

[50] G.F.R. Henderson, *Stonewall Jackson and the American Civil War* (New York: Barnes and Noble, Inc., 2006): pages 468-469. Henderson also claims that every move Jackson made was "actuated by some definite purpose."

secured Beverly's Ford after nearly a full day of fighting. On 22 August Jackson's command continued its pursuit of Pope and crossed the Hazel River at Wellford's Ford. On 25 August Jackson moved his army behind Pope near the town of Amissville, Virginia. He crossed the Hedgeman River at Henson's Mill, advanced to Orleans, and camped that night in the vicinity of Salem, Virginia.

Jackson's march around Pope on 25 August was one of the key moves in the entire campaign. Although the troops were exhausted, Jackson managed to position his command between Pope and Washington City, severing Pope's lines of communication with the U.S. capitol and interrupting the vital railways to Manassas, Virginia. Pope moved his army very cautiously because he was concerned with both Jackson and Stuart. Jackson had the luxury of advancing under the shield of Stuart's cavalry and was allowed to maneuver his army to force Pope to concentrate at Manassas. Speed was the main ingredient to Confederate success and Jackson, as usual, simply moved his army much quicker than Pope. Jackson's initiative allowed him to choose the battleground.

The Confederate march continued on 26 August and the command moved to Gainesville, Virginia via the Thoroughfare Gap in the Bull Run Mountains. Jackson then moved to Briscoe Station, along the Orange and Alexandria Railroad. Pope sat still and Jackson remained between Washington City and Pope's army. On 27 August Jackson moved to Manassas Junction to secure supply stores that Pope had left only three weeks earlier.[51]

Federal artillery positioned near Groveton on the Warrenton Turnpike opened fire on Third Brigade as it secured the stores. The men were proud to gain such a stockpile of fresh food and water. When Third Brigade left the Culpeper area two weeks earlier their supply wagons were not adequate and they were forced to live off the land. As they marched north, the men of Third Brigade ate unsalted raw corn, which caused many men to slow down because of weak stomachs. When the fresh food was secured at Manassas the men stuffed their haversacks "to the bursting point of every stitch in them."[52]

Jackson concentrated his command along an unfinished railroad, near the John Brawner Farm, on the evening of 28 August. Under the protection of a massive artillery barrage Jackson's brigades formed a line of battle that extended over 1,300 yards east from the Brawner Farm.[53]

The Forty-eighth Alabama, led by Lieutenant Colonel Abner Hughes in Sheffield's absence, entered the battle. Hughes commanded the regiment until 18 September. As the fight at Brawner Farm intensified, Jackson ordered Major John Pelham to

---

[51] OR, S1, V12, P2, page 184, T.J. Jackson's report.

[52] Huffman, *Ups and Downs of a Confederate Soldier*, pages 59-60.

[53] Hennessy, *Return to Bull Run*, page 178.

move his horse artillery to the Brawner orchard, with Colonel W.S.H. Baylor's "Stonewall Brigade", which was heavily engaged with the Nineteenth Indiana of the infamous "Iron Brigade". Although they had been surprised by the sunset attack and forced to form a line of battle, the Nineteenth Indiana held their ground against Pelham and Baylor very effectively.[54] The Stonewall Brigade, firing from behind a fence, attempted two bayonet charges and were repulsed both times by the Hoosiers. One Indiana private described the fighting as being "like two dogs fighting with a fence between them". Another described both sides as fighting "like demons".[55] Henry F. Young, a private in the Second Wisconsin, recalled the ferocity of the confederates: "…Jackson's men fight like Devils!"[56]

Taliaferro's Third Brigade remained out of sight in a wooded area near the northern edge of the Brawner Farm. It crossed Young's Creek, and marched to the Brawner orchard to cover the right (western) flank of the Stonewall Brigade. There, the brigade

---

[54] Alan T. Nolan. *The Iron Brigade: A Military History* (New York: MacMillan, 1961): pages 81-89. The brigade in which the Nineteenth Indiana served was infamous for all its regiments wearing similar black hats and were known simply as "The Black Hat Brigade". The brigade did not earn its nickname "The Iron Brigade" until the Battle of South Mountain in September 1862. Members of the Nineteenth Indiana remembered Taliaferro's entrance into the battle from the woods as being "dramatic".

[55] Alan D. Gaff. *On Many a Bloody Field: Four Years in the Iron Brigade* (Indianapolis: Indiana University Press, 1996): pages 156-157.

[56] Jeffry D. Wert. *A Brotherhood of Valor: The Common Soldiers of the Stonewall Brigade, C.S.A., and the Iron Brigade, U.S.A.* (New York: Simon and Schuster, 1999): page 150.

engaged the left flank of the Nineteenth Indiana. Taliaferro's position outflanked the Nineteenth Indiana, and the Alabamians and Virginians poured murderous musket volleys into the Hoosiers' ranks.[57]

The Forty-eighth Alabama, in the rear of the brigade, guarding wagons, prisoners, and cattle, marched one mile to join Third Brigade and took position in line of battle along the embankment of the unfinished railroad. Heavy fighting ensued, and the Forty-eighth Alabama, along with the rest of Third Brigade, made three charges toward their enemy's lines. Each charge ended with the same result, and the regiment fell back to the defenses of the railroad embankment. Sunset finally ended the fighting.[58]

The twilight engagement was spectacular. William Calvin Oates, colonel of the Fifteenth Alabama Infantry Regiment of Issac Trimble's brigade, was engaged on the Confederate left (east) flank. He described the scene: "…everything around was lighted up by the blaze of musketry and explosion of balls like a continuous bright flash of lightning."[59] After dark, the Nineteenth Indiana began to

---

[57] Hennessy, *Return to Bull Run*, page 178.

[58] Augustine Woodliff letters circa 1890. Woodliff became captain of Company G, Forty-eighth Alabama upon the resignation of Captain John S. Moragne on August 15, 1862. He commanded the company until he resigned on September 23, 1862. Woodliff suffered, as many Civil War soldiers did, from chronic diarrhea. Copies of the letters are in possession of the author.

[59] Oates, *The War Between the Union and the Confederacy*, pages 138-143.

fall back. John Gibbon, commanding the brigade, later described the slugfest between the two brigades as "…the most terrific musketry fire I ever listened to."[60] Gibbon assumed the battle was over, and at 9:00 p.m. ordered the brigade to fall back to the cover of the Groveton Woods. Taliaferro attempted to pursue Gibbon, but the brigade did not advance farther than the Brawner farmhouse.[61]

On 30 August the Forty-eighth Alabama continued fighting at Manassas. Starting at 3:00 p.m., each man of the Forty-eighth Alabama, along with all the men of Third Brigade, fired forty rounds into the Iron Brigade and charged until within forty yards of the Federal lines, which were four rows deep. Gibbon's Brigade fell back away the Brawner Farm during the afternoon. As the Forty-eighth Alabama pursued Gibbon's men it captured eight pieces of artillery.[62] The Federals did not, however, retreat from the Groveton Woods until Longstreet arrived on the brigade's left (western) flank. Gibbons retreated to the Dogan Ridge to regroup, and then returned to attack Jackson once again at the Groveton Woods.[63]

The Forty-eighth Alabama suffered fifty wounded at the

---

[60] James I. Robertson, Jr. *The Stonewall Brigade* (Baton Rouge: The Louisiana State University Press, 1963): page 146. Dr. Robertson used the quote from John Gibbon's *Personal Recollections of the Civil War* (New York, 1928).

[61] Nolan, *The Iron Brigade*, page 91.

[62] Woodliff letter circa 1890.

[63] Wert, *Brotherhood of Valor*, page 157.

Second Battle of Manassas.[64] Captain Moses Lee, the most capable and influential officer of Company K, was among the killed. The Jacksonville, Alabama native who had assembled Company K, was killed on August 30. Captain Alvin Oscar Dickson remembered the efforts of Captain Lee: "…he led the regt in a charge at Manases and drove the enemy in confusion and caused the enemy to give way at evry point – tho the galent capt went down to rise no more tho he still lives in the memory of those that new him."[65] Sheffield described him as "brave as Julius Caesar".[66] Lee was the first officer of the Forty-eighth Alabama killed in action.

It is interesting to observe the procedures of the Confederate Medical Corps during the Second Battle of Manassas. Each brigade had its own field hospital, headed by the Senior Surgeon of Brigade. The regimental surgeons were often near the front lines of their respective regiments and would perform emergency procedures on wounded soldiers whose injuries were not life threatening. Regimental surgeons were responsible for directing seriously wounded soldiers to the Senior Surgeon of Brigade. This system evolved into an effective field ambulance system. The "infirmary corps" was a group of soldiers that worked in the field removing the

---

[64] *OR, S1, V12, P2,* page 561. The number of killed is not known.

[65] *Dickson to Owen,* 8 October 1913. A marker of the traditional military style stands in the Jacksonville, Alabama city cemetery memorializing Captain Lee.

[66] *Jacksonville Republican,* August 28, 1862.

wounded. Assigned by the regimental surgeon, they were often equipped only with a revolver and a badge.[67] Doctors worked all night tending to the wounded. Amputations were often performed by candlelight with little or no chloroform. Men lay strewn around the infirmary "as thick as a drove of hogs in a lot" moaning in agony – begging for water and, in some cases, death.[68]

The struggle between Pope and Jackson came to an end only when General James Longstreet arrived to reinforce Jackson. Pope's raid into Virginia came to an abrupt halt at the battles of Cedar Run and Second Manassas. Robert E. Lee took command of the Confederate Army in June 1862. By the end of August 1862 it had thwarted two major invasions of Virginia by Union troops. Lee decided to take the war into the north in hopes that a major victory on northern soil might bring Great Britain and France into the war on the Confederate side.[69]

Jackson's command crossed the Potomac River into Maryland on 5 September at White's Ford to the tune of "Maryland, My

---

[67] Horace H. Cunningham. *Field Medical Services at the Battles of Manassas* (Athens: University of Georgia Press): 1968, page 75.

[68] Spencer Glasgow Welch, *A Confederate Surgeon's Letters to his Wife* (New York: The Neale Publishing Company, 1911): pages 26-27. Welch notes that most chloroform was saved for wounded officers.

[69] James M. McPherson, *Crossroads of Freedom: Antietam, The Battle that Changed the Course of the Civil War* (New York: Oxford University Press, 2002): pages 93-95.

Maryland". His army camped nearby at Three Springs, Maryland,[70] and the next day moved to Frederick City, Maryland. Camped near that town, the command remained until it was ordered to engage Federal troops and capture supplies near Harper's Ferry, Maryland and Martinsburg, Maryland. On 10 September, rather than marching straight to Martinsburg, Jackson moved his command back into Virginia via Middletown, Boonsboro, then to Williamsport, Virginia. He re-crossed the Potomac River at Light's Ford on 11 September. His army then moved to the North Mountain Depot, located seven miles northwest of Martinsburg along the Baltimore and Ohio Railroad. On 12 September most of Jackson's command, led by the cavalry, moved into Martinsburg without any Federal resistance and captured many supplies, including ordnance and much needed food. Jackson marched his troops to Harper's Ferry on 13 September, and before noon, engaged the Federal troops garrisoned in and around the arsenal.[71] Taliaferro's Third Brigade was not involved in the fighting. Jones' command was placed north of Harper's Ferry on the west bank of the Potomac River, on the cliffs north of the Baltimore-Ohio Railroad just north of the Old Furnace, near Captain William Pogue's artillery.[72] Jackson's attack on Harper's Ferry was a major success for the Confederate Army. The Federal Arsenal there

---

[70] Huffman, *Ups and Downs of a Confederate Soldier*, page 64.
[71] *OR, S1, V19*, pages 952-958, T.J. Jackson's report.

[72] Thomas Yoseloff. *The Official Atlas of the Civil War* (New York, 1958): plate 58-3.

yielded over thirteen thousand small arms, seventy-three pieces of artillery, and eleven thousand Confederate prisoners of war. Jackson's re-equipped army continued its march north.[73]

On 14 September Jackson's command moved near the Potomac River at Bolivar Heights, Maryland. Bolivar Heights guarded the ford where Jackson moved his army across the Potomac River. Federal skirmishers engaged the approaching Confederates but were quickly dispersed by Confederate artillery. At dawn 15 September the remaining Federal troops near Harper's Ferry escaped to join the main Federal army. Jackson anticipated a major concentration of armies, and at 3:00 p.m. he gave orders to cook two days rations. Jackson's command spent the next ten hours preparing for the next march. At 1:00 a.m. on 16 September, Jackson's command resumed its march north into Maryland. It reached the Potomac River at dawn and continued to the northern vicinity of Sharpsburg, Maryland, to a farm owned by Alfred Poffenberger west of the Hagerstown Pike.[74]

In the late afternoon of 16 September Brigadier General John R. Jones, division commander, ordered the formation of a double-line (two brigades in front of the other two brigades) and awaited further orders. The right of Jones' line stood to the left of Brigadier General John Bell Hood's brigade along the Hagerstown Pike into the West

---

[73] *OR, S1, V19*, pages 952-958, T.J. Jackson's report.

[74] *OR, S1, V19*, pages 1006-1008, Jones' report.

Woods. Third Brigade also had a new commander. Colonel A.G. Taliaferro was wounded at Second Manassas and replaced by Colonel James W. Jackson of the Forty-seventh Alabama. Third Brigade, one of the rear brigades of Jones' double-line, followed the rest of the brigade into the West Woods against Federal infantry there. During the early morning of 17 September the division immediately became targets to heavy artillery fire from the northeast. Jones reported the Federal guns blasted into them, "…a storm of shell and grape…," until the entire division was forced to fall back. After regrouping his division, Jones prepared the men for another assault on the Federal lines near the West Woods.[75]

Positioned approximately six hundred yards north of the Dunker Church along the Hagerstown Turnpike, Colonel J.W. Jackson's (A.G. Taliaferro's) Third Brigade advanced toward the Miller Cornfield where it encountered an old foe – Gibbon's "Iron Brigade". Third Brigade was ordered to lie down and await the oncoming Iron Brigade. When ordered to rise and face their enemy, Third Brigade poured a murderous volley into the Midwesterners, who were unaware of the Confederates' presence. A bloody fight ensued between Third Brigade and the Iron Brigade. Third Brigade held its own until it became targeted, yet again, by Federal artillery. The brigade was forced to fall back across the Hagerstown Turnpike

---

[75] *OR, S1, V19,* pages 1006-1008, Jones' report. It is ironic that Colonel Jackson took command of Third Brigade, especially after the debacle at Cedar Run a month earlier.

into the West Woods. The Iron Brigade, according to Major Williams, also retreated, "...unable to withstand the resolute valor of our troops." After its engagement with the Iron Brigade, Third Brigade fell back as far as Hauser's Ridge. After taking heavy casualties in the West Woods along the Hagerstown Turnpike, Third Brigade was knocked out of the Battle of Sharpsburg early, by 9:30 a.m.[76]

In its engagements at Sharpsburg on 17 September, the Forty-eighth Alabama Infantry Regiment, one of four regiments of Third Brigade, suffered a total of forty two casualties. Captain R.C. Golightly of Company H, led the fight from the front was shot and killed instantly. He was the only officer of the Forty-eighth Alabama killed at Sharpsburg, but nine enlisted men, including Captain Golightly's brother Thomas, died. Four officers and twenty-eight men were wounded.[77] Private John M. Anderson of Company K wrote that the regiment was, "...verry badly cut up and destrored," with "greate destruction among our por soldiers."[78]

---

[76] *OR, S1, V19,* pages 1012-1013, Williams' report.

[77] *OR, S1, V19,* page 1009. John Dykes Taylor placed Golightly at the front of the regiment at Sharpsburg (see John Dykes Taylor, page 12); Captain Golightly was replaced by William "Mack" Hardwick (see Oates' *War*, page 795).

[78] *John M. Anderson to Elizabeth Anderson,* 28 September 1862. Anderson was absent during the Sharpsburg campaign due to illness. He received information from soldiers of the Forty-eighth Alabama as they entered the hospital at which he was being treated in Lynchburg, Virginia.

Jackson's command remained in battle formation at Sharpsburg on 18 September. The following day it crossed the Potomac River at Shepherdstown and remained there as reserves for two days. It moved toward Martinsburg, and the Division camped at Opequon. The men remained there until it moved to Bunker Hill, Virginia on 28 September.[79]

After the Battle of Sharpsburg, Lee's army was forced to do something it had not done under his command – retreat. 17 September 1862 stands as the bloodiest single day in American history as over thirty thousand men fell killed or wounded. After the battle, the Army of the Potomac boasted of complete victory, and for the first time was confident and high in morale. Historian James McPherson argued that although the Army of Northern Virginia lived to fight another day it was never the same. Lee did not wish to leave Maryland to McClellan, and actually wanted to re-cross the Potomac River and move back into Maryland, but he realized that his army was too tired and weakened. Politically, the battle crippled the Confederacy. The possibility of European support for the Confederacy ended, and Abraham Lincoln issued his Emancipation Proclamation. The Battle of Sharpsburg was, indeed, the beginning of the end for the Confederate States of America.[80]

---

[79] *OR, S1, V19*, page 1006, T.J. Jackson's report; see also *OR, S1, V19*, page 1015, Stafford's report.

[80] McPherson, *Crossroads of Freedom*, pages 133-156.

Joshua Price

3

FREDERICKSBURG TO SUFFOLK

"I would have been proud of the priviledge of dying on such a field."
- Captain Thomas Eubanks, Company D, 48th Alabama

By December 1862 the weather was cold and wet. Third Brigade, along with the rest of Jackson's Corps, was on the move. During the march from Winchester, Colonel Sheffield, who had returned to command the regiment on September 18, was notified that many men in the Forty-eighth Alabama, like most men in the Confederate army, were barefooted and many could no longer walk. Many were being moved by ambulances.[81] Those who managed to march tied old rags around their naked feet to protect against the frozen ground. "Sometimes a man's foot would freeze and he would lose some of his toes or a part of his foot. Then he could no longer fight."[82] While meeting with Colonel Warren, commanding Third Brigade, Sheffield threatened to halt the regiment's march and make camp where it was. He would wait until spring rather than march his men on frozen ground without shoes. Warren could not supply Sheffield's regiment with new shoes, but he did accommodate Sheffield with enough empty wagons to haul the men. The regiment, along with

---

[81] Hoole, *John Dykes Taylor*, page 13.

[82] Andrew Davidson Long, *Stonewall Jackson's Foot Cavalryman* (Austin, Texas, 1965): page 14.

Third Brigade, moved into Guinea's Station, south of Fredericksburg, where it received more shoes to resume the march.[83]

On the evening of 11 December W.B. Taliaferro received orders to march north toward the town of Fredericksburg, Virginia and prepare for battle. At dawn the next day, he began marching his men from their encampments at Guinea Station toward Fredericksburg. After a few hours the division was placed in battle formation on Jackson's right at Hamilton's Crossing south of Fredericksburg. The Hamilton House stood at the crossing, located west of the Richmond, Fredericksburg, and Potomac Railroad near the Military Road. With a commanding view of the Federal lines, Jackson ordered artillery placed there, and Warren was ordered to move his command to the rear of the Hamilton House and guard the artillery.

On Saturday, 13 December, Third Brigade was relieved of its position by Jubal Early's command. Third Brigade was positioned behind Paxton's brigade – with Sheffield and the Forty-eighth Alabama posted at the brigade's right flank. At approximately 10:00 a.m., the thick fog cleared and the Battle of Fredericksburg began. Confederate artillery blasted the Union troops as they attempted to cross the frozen fields that lay in front of Jackson's defenses. Third Brigade advanced as far as the second line but did not engage in combat. It was ordered to fall back to the Military Road and occupy it. The Forty-eighth Alabama, either from skirmishing or by enemy

---

[83] Hoole, *John Dykes Taylor*, page 13.

artillery, suffered only five casualties at the Battle of Fredericksburg.

The day after the sanguine battle, Third Brigade was ordered to move to Paxton's left, along the front line, but saw little combat except for minor skirmishing. The Forty-eighth Alabama lay in a ditch all day and night, cold and wet, waiting to attack the Union lines if ordered to do so. That night, the regiment watched in awe of the stars in the skies. In a letter home, John M. Anderson of Co. K, cheerfully asked his wife, "Did you see them great lites and sighns in the elements last Sunday [December 14] night? I saw them when we was in that ditch on line of battle."[84] After only a day on the front line, Taliaferro's Division was ordered to the rear of Jackson's line. At dawn on 15 December, Taliaferro's Division was relieved by D.H. Hill's command, and it retired to the Mine Road.[85]

The weather was terrible. Although the exact temperature is not known, it is likely that the Virginia winter was well below freezing as ice, rain, and snow combined to torment the soldiers at Fredericksburg. Soldiers without tents slept on the ground and awoke each morning to frost on their blankets.[86]

---

[84] *John M. Anderson letter to Elizabeth P. Anderson*, 19 December 1862.

[85] OR, S1, V21, pp. 675-677, W.B. Taliaferro's report and also Warren's report, *ibid.* pages 685-686.

[86] George Rable, *Fredericksburg! Fredericksburg!* (Chapel Hill: University of North Carolina Press, 2002): pages 91, 100-101. Rable sums up the entire combat experience of Third Brigade at Fredericksburg in one sentence in his endnotes; John Dykes Taylor also reported that the Forty-eighth Alabama saw no combat at

Jackson's Corps marched south from Fredericksburg on 17 December. The Forty-eighth Alabama, along with Taliaferro's (Jackson's) Division, made its winter camp of 1862-63 at Moss Neck, Virginia south of the Rappahannock River.[87] Jackson ordered pickets to watch the Rappahannock River continuously in case of an unexpected Federal offensive. This task required one full brigade per week. The unfortunate brigade was randomly selected, removed from the comforts of their camp, and moved north to a new camp where the men picketed around the clock for a full week. Another brigade would be ordered to relieve it. The Forty-eighth Alabama was ordered to picket for a week in late December. Anderson of Company K spent Christmas 1862 in a rifle pit above the Rappahannock River "ketching and killing 'Confederate bodyguards' (lice)". While on picket duty, many men conversed with the Union soldiers on picket duty across the river. Many of the Confederates even traded tobacco for coffee and sugar. It was a peaceful, but miserable week.[88]

While at winter camp, brigades had daily drill, roll calls, and inspections. The men made huts out of anything they could find. Most men used logs, but some men used their tents, if they still had them. Most of the men did not have shoes. To improvise, they were

---

Fredericksburg; see *OR, S1, V21*, p. 562 for casualties of the Forty-eighth Alabama at Fredericksburg.

[87] Hoole, *John Dykes Taylor*, page 13.

[88] *John M. Anderson letter to Elizabeth P. Anderson*, December 28, 1862.

ordered to use the hides of slaughtered cattle to make moccasins. Because of the shoe problem within Jackson's division, a shoemakers division evolved. To pass the time during winter camp, the men played games such as marbles, chess, and cards. One of the camp favorites was "snowballing". For entertainment, brigades would engage other brigades in snowball fights. It must have been quite a scene![89]

Even during the "lax" winter quarters, Confederate camps did not go without regular military discipline. Many men tried to desert from the ranks. The men who were caught were always given a trial by a court-martial of either brigade or regimental officers. If found guilty, the men were subjected to a variety of punishments, including being forced to wear a "barrel shirt", a ball and chain, or being "bucked and gagged". A barrel shirt was an empty keg with holes cut in it for the arms, legs, and head. If found guilty of desertion, the "barrel shirt" would have the word "deserter" written on it. The man would be marched continuously for many hours around the camp for all to see. If a man were sentenced to be "bucked" he would be forced to sit with his knees against his chest. A rifle or stick was placed underneath his knees, and his wrists were tied together underneath the rifle. If the man were to be "gagged" as well as "bucked", a bayonet or stick would be placed in his mouth

---

[89] John H. Worsham, *One of Jackson's Foot Cavalry* (New York: The Neale Publishing Co., 1912): pages 151, 155-156; for more on snowball fighting, see also William Stanley Hoole, ed., "The Letters of Captain Joab Goodson, 1862-1864," *Alabama Review*, April 1957, pages 138-139.

and tied to the back of his neck. The punishment usually lasted from sunrise to sunset. This was one of the most dreaded forms of punishment within the Army of Northern Virginia.[90] Turner Vaughan witnessed a deserter from the Forty-eighth Alabama paraded around the camp. He wore a "barrel shirt" with the word "deserter" on it, and a ball and chain tied to his ankle. A band of musicians led him.[91] The most unfortunate soldiers guilty of desertion were sentenced to death by firing squad. In the case of this capital punishment, the entire brigade was assembled to witness the event. The condemned were always marched in front of their comrades, the entire division, before being taken to the place of execution. In one instance, a young man condemned to death was marched to an open grave, where he sat upon his coffin with his hands tied. A squad of twelve soldiers was marched to within six feet of him and was ordered to fire into his chest when given the command.[92]

Minor infringements were not overlooked. Men received various forms of punishment for drunkenness, absence without leave (AWOL), theft, and other crimes. These were punishable by painful sentences similar to that of desertion. For drunkenness, the guilty

---

[90] Casler, *Four Years in the Stonewall Brigade*, pages 100-101.

[91] Turner Vaughan, "The Diary of Turner Vaughan, Company C., Fourth Alabama Regiment, C.S.A.," *Alabama Historical Quarterly 18*, no.4 (December 1956): page 593.

[92] Spencer Glasgow Welch, *A Confederate Surgeon's Letters to his Wife* (New York: The Neal Publishing Co., 1911): pages 44-45.

man often wore the barrel shirt with the word "DRUNK" written on it. If found AWOL, the man would be forced to carry a heavy rail, usually a railroad crosstie, for eight hours under guard. If guilty of theft, the man would either be "bucked" or have his thumbs tied together over a tree branch. There he would stand on his tiptoes, often from sunrise until well after sunset.[93] All punishments were carried out swiftly and effectively, according to Jackson's regulations.

The Forty-eighth Alabama had served under General Thomas Jackson since it arrived in Richmond in June 1862. It had participated in two major campaigns and four major battles. The regiment had suffered casualties from battle, fatigue, and disease, but the men revered their commander. "But it did not matter how hungry we were, how barefooted or how dirty – we always cheered when Jackson came along on 'Old Sorrel'."[94]

On 19 January 1863 the Forty-eighth Alabama, along with the Forty-seventh Alabama, was transferred out of Taliaferro's Third Brigade and into Evander McIver Law's Brigade. Law's Brigade was placed in John Bell Hood's division of James Longstreet's First Corps. Law's restructured brigade consisted of five Alabama regiments: the Fourth, the Fifteenth, the Forty-fourth, the Forty-seventh, and the Forty-eighth.

---

[93] Casler, *Four Years in the Stonewall Brigade*, pages 100-101.

[94] Long, *Stonewall Jackson's Foot Cavalryman*, page 15.

The Forty-eighth Alabama remained camped near Guinea Station until it was ordered to move to Richmond on 18 February 1863. It arrived two days later after having marched through two snowstorms.[95] On 1 March it received orders to march back to Fredericksburg. After sixteen miles, the brigade bivouacked near the town of Ashland, Virginia. The weary soldiers, many without shoes, made camp in the freezing snowstorm. The men quickly built fires, wrapped themselves in their blankets, and rested as best they could. The following morning they woke to find a half foot of snow layered all over them. Then the brigade received orders to march back to Richmond. The tired and frustrated brigade re-trekked its route of the previous day. "There was more suffering on those sixteen miles than on any other of the same length during our experiences as soldiers. The barefooted, illy shod…left a trail of bloody footprints behind them."[96] The Forty-eighth Alabama, along with the rest of Law's Brigade, remained at its encampment at Richmond until it was marched to Suffolk, Virginia in mid-April 1863 to participate in the siege of that town.[97]

No reports were submitted by officers of the Forty-eighth Alabama describing the events of the Siege of Suffolk. However, Captain Thomas Eubanks of Company D wrote a letter home to his friend describing in detail the experience of the Forty-eighth

---

[95] Hoole, *John Dykes Taylor*, pages 14-15.

[96] Stocker, *From Huntsville to Appomattox*, pages 89-90.

[97] Hoole, *John Dykes Taylor*, pages 14-15.

Alabama at Suffolk:

"I wrote you a few lines about ten days ago which I hope you received, in which I stated I would soon write you again. I commenced to write to you yesterday but was so interrupted that I burned it. We were under furious fire yesterday and for near ten days past. On day before yesterday the Yankees attacked a fort of ours and took it with two companies of the 44th Ala. On yesterday evening Gen'l Hood called for a hundred men from each brigade as volunteers to retake the fort – making three hundred men – they were warned of the hazardous undertaking. The men volunteered and were placed under command of a Capt. from this brigade. I first volunteered to take command of the men from this regt. I then offered my services at Brigade Headquarters where they were accepted – I was then honored with the command of all the volunteers from this brigade, being right of the line. We moved off about dark to make a night attack on the fort but; low and behold, when we got near it we met Gen'l Hood who to the mortification of many of us, informed Capt. Couzins that he had postponed the attack until tonight. We awoke this morning and found the Yankees had evacuated the fort. To tell you the truth my "Good Friend", I would have been proud of the priviledge of dying on such a field engaged as I would have been, in command of one hundred men, who volunteered to die. In one night we would have been victorious and gloriously so. I will tell you that on Monday the 13th first I had the honor to head one of Gen'l Hood's expeditions consisting of three companies of this regt. The order was to advance our companies until we drew fire of the enemy. There were two Capts. and one Major present. I placed myself twenty paces in front of our line to command foreward, the men and officers followed, the enemy immediately opened with round shot, shell, grape, and canister, but still we advanced. I led until every officer refused to go further for the reason that we had fulfilled the order. We were very near the Yankee fortification and had the men and officers followed I should have led them to where they could have picked off Yankees with muskets. I then asked for three men to volunteer and we went to the creek in front of their works and took a few fires and went back to our commands."[98]

The regiment did not leave Suffolk until the town was secured by the Confederate army on 3 May 1863. During the night, the regiment was ordered to march back to Richmond, then to Culpeper Court House. It camped along the Rapidan River, where the men received much needed new clothing. Lee ordered the brigade to rejoin the main body of the army triumphant after his great victory at the Battle of Chancellorsville.[99]

---

[98] *Thomas J. Eubanks to Mrs. Foster*, 21 April 1863. Copy of letter in appendix.

[99] Hoole, *John Dykes Taylor*, pages 14-15.

4

NORTH TO GETTYSBURG

"We fought the enemy until night put an end to the day's work of death."
- Sergeant John Dykes Taylor, Company D, 48th Alabama

By late April 1863, a major clash between the Army of Northern Virginia and the Army of the Potomac was imminent. "May God have mercy on General Lee, for I will have none" had been "Fighting Joe" Hooker's promise to Lincoln. Hooker did not live up to his promise to the President. By 6 May 1863 the Army of Northern Virginia, without Longstreet and two of his divisions, had routed the Union army at Chancellorsville, and Hooker had pulled back toward Washington City. Lee was poised to deliver the *coup de grace* upon the Army of the Potomac.[100]

After the death of Thomas J. "Stonewall" Jackson on 10 May 1863, Lee decided to restructure his seemingly invincible army by adding a third corps. To accomplish this, Lee removed divisions and brigades from the two existing corps and placed them into the newly

---

[100] James M. McPherson, *Battle Cry of Freedom* (New York: Oxford University Press, 1988): 639-645. Prior to the Battle of Chancellorsville General James Longstreet, with John Bell Hood's and George Pickett's Divisions, had been detached south to Suffolk, Virginia to attack the enemy at that town and also to seek surplus supplies. They did not rejoin Lee's main army until 3 May 1863.

created Third Corps. James Longstreet, Richard Ewell, and Powell Hill commanded Lee's three corps.

By 1 June 1863, the reorganized Army of Northern Virginia was rested and ready to fight again. Lee decided that the war should be taken north into Pennsylvania to relieve Virginia of the destruction the war had caused over the previous two Aprils. If Lee could defeat his enemy in Pennsylvania, Lincoln might be forced to consider peace negotiations. Also, a victory in the eastern theater could counter the imminent loss of the Mississippi River to Ulysses S. Grant's army at Vicksburg. On 4 June, the lead elements of Longstreet's First Corps began the march north through the Shenandoah Valley. The Forty-eighth Alabama Infantry Regiment, attached to Brigadier General Evander McIver Law's Brigade of Major General John Bell Hood's Division of the First Corps, followed.

On the first day of marching, the regiment crossed the Rapidan River at Raccoon Ford and marched to Culpeper Court House, Virginia, where it camped and drilled for several days. There, the regiment witnessed the great cavalry review of Stuart's command. The entire brigade then marched past the old Cedar Run battlefield of 1862, where the soldiers discovered the exposed remains of some of their comrades killed less than a year earlier. The regiment stopped to rebury their fallen comrades, then it marched toward Maryland on 14 June. It marched uninterrupted from Cedar Run to Ashby's Gap. The next day the regiment marched through

Martinsburg, Maryland and camped that night near the Potomac River at Williamsport. There, the men enjoyed much needed rest and relaxation. The regiment stayed there until 26 June when it crossed the Potomac River. The following morning the regiment marched toward Greencastle and bivouacked a few miles inside Pennsylvania.

The Forty-eighth Alabama, with Law's entire brigade, arrived in Chambersburg, Pennsylvania on 29 June with the tune of "Dixie" from the regimental strings. The citizens of Chambersburg were eager to get a glimpse of the infamous Army of Northern Virginia. They lined the streets "almost to suffocation" in their finest clothes, waving the flag of the Union. Many of the citizens even voiced a few unwelcoming remarks to the passing soldiers.[101] In one particular instance, a lady wrapped only in a Union flag which nearly exposed her breasts, taunted and insulted the ragged Alabama and Texas soldiers as they marched under her balcony. After ignoring her for some time, one rowdy Texan shouted to her, "Take care, madam, for Hood's boys are great at storming breastworks when the Yankee colors is on them!" To this the men cheered and the lady stormed furiously back into her quarters.[102]

---

[101] *John Dykes Taylor*, page 17.

[102] Walter Lord, ed., *The Fremantle diary: being the journal of Lieutenant Colonel James Arthur Lyon Fremantle, Coldstream Guards, on his three months in the Southern States* (Boston: Little Brown Publishing Company, 1954): page 191; Henry Figures mentioned in his 18 July 1863 letter that most of the

The march to Pennsylvania had been relatively pleasant and the brigade's morale was high. Perhaps to divert their minds from the hardships of marching, many of the men gazed upon the natural beauties of the Pennsylvania landscape, reminiscent of north Alabama. The Forty-eighth Alabama's newly appointed adjutant, Lieutenant Henry Stokes Figures of Huntsville, Alabama, often thought of "dear old home" as the army moved north.[103]

Although there were many instances of Confederate soldiers looting farms in Pennsylvania for food and supplies, the men of the Forty-eighth maintained their best conduct. No known complaints were made by the Pennsylvanians against soldiers of the Forty-eighth, and no instances of misconduct were reported. The first night in Chambersburg the soldiers were allowed to burn rails and cordwood for heat. "You never saw fences disappear as fast as they did that night" recalled Figures. On the morning of 30 June, the regiment marched to Fayetteville, approximately five miles east of Chambersburg and twenty-two miles west of Gettysburg. The regiment established camp and remained there until 2:00 a.m. on 2 July when it was ordered to march toward the fighting at

---

Confederate soldiers were well mannered until the Chambersburg, Pennsylvania where the women began insulting them with foul language.

[103] *Henry Stokes Figures to his sister*, July 18, 1863. Figures was Ordnance Sergeant of the Fourth Alabama Infantry from April 1861 until his transfer and promotion to Adjutant of the Forty-eighth Alabama in the spring 1863. Figures' thirty-one letters, dating from 10 April 1861 to 29 February 1864, are the best primary source known to exist of everyday life and battle experiences of the officers and men of the Forty-eight Alabama Infantry Regiment. The historian of Gettysburg National Military Park provided the author with copies of the letters.

Gettysburg.[104]

Law's Brigade arrived on the battlefield at Gettysburg at approximately 2:00 p.m. on 2 July after the men marched over twenty miles with little or no rest. Major General John Bell Hood's Division, to which Law's Brigade was attached, was ordered to the southern end of the battlefield facing the rocky heights that would be known as Big Round Top and Little Round Top. The brigade was tired and hungry from its march, and the men rested only a few minutes before forming into battle-line. Due to fatigue resulting from the forced march these soldiers were in no condition to conduct a major assault.

Hood's Division formed into two brigade lines on the southern end of Seminary Ridge. Of Hood's four brigades, Law's was positioned on the right of the frontline. The Alabamians, approximately 1,500 strong, stood as the extreme right of the Confederate Army – they were *the* end of Lee's gray line. The battle order of the regiments, from left (north) to right (south), were as follows: the Fourth Alabama, the Forty-seventh Alabama, the Fifteenth Alabama, the Forty-fourth Alabama, and the Forty-eighth Alabama, holding the brigade's southernmost flank.[105]

Longstreet's orders to Hood were simple. Hood was to

---

[104] *Figures letter to sister*, 18 July 1863.

[105] Harry Pfanz, *Gettysburg: The Second Day* (Chapel Hill: The University of North Carolina Press, 1987): 159-161, 170-171.

approach the Round Tops and disperse their enemy from its positions by way of frontal assault. Hood and his brigadiers, including Law, did not approve of the orders and protested them many times. Three times Hood begged Longstreet to allow him to use the terrain and march behind the enemy by way of Big Round Top to attack from the south. Longstreet denied the requests and insisted that Lee's orders for a frontal assault be carried out immediately. Nobody truly knew the strength of the Federal lines upon Little Round Top nor recognized the effect that the rocky heights provided in defense. At 4:00 p.m. Hood, under protest, was ready to attack.[106]

When the attack began, Law's Brigade immediately came under heavy artillery fire from the rocky spur (called "Devil's Den") below Little Round Top. Also U.S. Sharpshooters fired from the base of Big Round Top at Plum Run. There was little that Hood could do to counter the artillery fire coming from Devil's Den, but he could remove the pesky snipers who lay in the grassy fields of the Slyder Farm. The Second United States Sharpshooters "did splendid execution" on the advancing Alabamians and temporarily halted one of the regiments. Before he turned it toward the Devil's Den, Hood ordered Reilly's Battery to barrage the fields where the sharpshooters were posted. Law then ordered Colonel James Sheffield to send two companies (A and H) from the Forty-eighth

---

[106] John Bell Hood, *Advance and Retreat: Personal Experiences in the United States and Confederate States Armies* (New Orleans: Hood Orphan Memorial Fund, 1880): pages 56-59.

Alabama as skirmishers to engage the sharpshooters and disperse them into the hills. The snipers were soon swept from the field, and the Alabamians continued their advance without the threat of Union snipers.[107]

These immediate obstacles did not slow the advance of Law's Brigade, but the federal artillery soon became a nuisance as the Alabamians were completely exposed in the open fields of the Slyder Farm. Suffering steadily increasing casualties, Law was now forced to make a critical decision. There was little doubt that the Federal battery in the Devil's Den must be eliminated in order for the attack to continue. That task was left to Law's troops. As the Confederates approached Plum Run, Law made the decision to silence the guns. He ordered two of his regiments toward the Devil's Den to attack the Federal battery. Likely for the sake of disguising his maneuver from the enemy, Law moved his two right-flank regiments, the Forty-eighth Alabama and the Forty-fourth Alabama, behind the rest of the brigade advancing into Plum Run Valley and the Devil's Den.[108]

---

[107] J. Gary Laine and Morris M. Penny, *Struggle for the Round Tops: Law's Alabama Brigade at the Battle of Gettysburg, July 2-3, 1863* (Shippensburg: White Mane Publishing Company, 1999): pages 30-34. According to Laine and Penny's calculations, the Forty-eighth Alabama could only muster 275 officers and men on the battlefield after the grueling fourteen hour march from New Guilford (page 34); *OR, S1, V27, P1*, pages 518-519, report of Major Homer R. Stoughton, Second U.S. Sharpshooters.

[108] Evander M. Law, "The Struggle For Round Top," *Battles and Leaders of the Civil War, Volume III* (New York: The Century Company, 1888): pages 318-326.

Colonel William F. Perry's Forty-fourth Alabama led the approach and took the initial brunt of the Federal barrage. The first Union volley made casualties of nearly one quarter of the soldiers. Still, the Forty-fourth charged and made quick work of the battery. It immediately became desperately engaged with the Fourth Maine Infantry in the Devil's Den. The Union batteries on Little Round Top summarily opened fire upon the Forty-fourth.[109]

The Forty-eighth Alabama, acting in support, followed the Forty-fourth Alabama into Plum Run Valley at approximately 4:30 p.m. As the Forty-fourth engaged the Fourth Maine, the Forty-eighth Alabama formed its battle line to the right of Perry's regiment. Sheffield's men were now in the valley between the Devil's Den and Little Round Top, and both Alabama regiments faced the Fourth Maine.

Colonel Sheffield deployed the Forty-eighth Alabama in a position to pressure the left flank of the Fourth Maine. Although the Fourth Maine had poured eight volleys into his ranks, Sheffield waited until his regiment was within twenty paces of the enemy before its first volley was fired.[110] The fighting that ensued was as fierce as any the regiment experienced during the entire war. The fight lasted over an hour and a half, and both sides suffered heavy casualties, but neither side lost its position. Bullets whizzed through the air and pinged off of the boulders that lay strewn in the Devil's Den. Many men fell wounded and many others fell dead. Private

---

[109] *OR, S1, V27, P2*, pages 393-394, William F. Perry.

[110] Laine and Penny, *Struggle For the Round Tops*, pages 56-57.

John M. Anderson of Company K recalled, "The shell and shot fell like hail among us while the miney balls was whistling around on every side. I thought I was doing well to get off the battlefield alive while many others was falling around me dead and dying."[111]

Meanwhile, Robertson's Texans poured over the southern end of Houck's Ridge and into the Devil's Den. The Federals posted atop Little Round Top began firing into the right flank of the Forty-eighth Alabama and forced it to retire to the rocks near the Devil's Den for protection. "We was in a warm place," Anderson later recalled.[112] Sheffield, understanding that the fight for the Devil's Den should be left to the Texans, turned his attentions toward Little Round Top. After falling back a short distance and reforming its lines, the Forty-eighth Alabama found itself supporting the efforts of the Fourth Texas Infantry Regiment in its attempts to drive the enemy from the formidable positions atop Little Round Top.[113] The Forty-eighth assaulted Little Round Top four times. While observing Sheffield's behavior in the heat of battle, Figures recalled, "I never saw a braver man in my life than Col. S. is." Figures carried the regimental battle flag, after its previous three bearers had been killed, during one of the assaults up Little Round Top.[114] Unaware

---

[111] *John M. Anderson letter to wife Elizabeth P. Anderson*, 13 July 1863.

[112] *John M. Anderson to Elizabeth P. Anderson*, 13 July 1863.

[113] *OR, S1, V27, P2*, pages 410-411, Sheffield.

[114] *Figures*, 8 July 1863; The *Jacksonville Republican* listed Third Sergeant James Madison Parrish of Company C as one of the color bearers killed while

of his promotion to brigade commander, Sheffield remained in the thick of the fight, leading the Forty-eighth Alabama from the front. John Dykes Taylor used adjectives such as "superhuman," "consummate," and "unflinching" to describe the courage and efforts made by the colonel during the struggle at Little Round Top. Sheffield even urged the Texans forward against "a galling fire" from the enemy. He told them "they were Texans and never to yield".[115]

During the assaults against the Devil's Den and Little Round Top, General Hood was seriously wounded by a piece of shrapnel in his left arm and was taken out of the battle. Hood's absence created confusion among the brigade and regimental commanders because they were not sure if their orders would remain the same. Law, the senior brigadier, immediately took command of the division. Sheffield, the highest ranking regimental commander, took command of Law's Brigade. Thus, the Forty-eighth Alabama, in the midst of attacking Little Round Top for the second time, lost its commander (Sheffield) to promotion. Captain Thomas Eubanks, of Company D, was now the ranking officer of the Forty-eighth Alabama after Lieutenant Colonel William "Mack" Hardwick and a Major Columbus B. St. John were wounded. Eubanks led the final two charges against Little Round Top. Despite the efforts of

---

assaulting Little Round Top. According to Parrish's military records he was, rather, mortally wounded with a "ball in side and lung" and died as a prisoner of war at Fort Delaware, New York on 26 July 1863 and is buried in the adjoining cemetery. The names of the other bearers are unknown.

[115] *John Dykes Taylor*, page 18.

Sheffield and Eubanks, the Confederates could not breach the Union entrenchments. The Forty-eighth, along with all the other attacking Confederate units, was violently repulsed.[116]

Sheffield's command of five regiments stretched from Devil's Den through Plum Run Valley, across the northern edge of Big Round Top and around the rocky heights of Little Round Top. There, Colonel Joshua Chamberlain's Twentieth Maine furiously fought Colonel William Oates' Fifteenth Alabama regiment. After nearly two hours of hard fighting the exhausted regiments began to fall back to the wooded shelter below Big Round Top. They regrouped and considered another assault, but nothing further happened. Taylor recalled that the regiment "fought the enemy until night put an end to the day's work of death." The entire brigade, what was left of it, fell back during the night and formed a new line, from north to south, from Devil's Den to the southwestern base of Big Round Top. During the night, snipers from both sides assured that their foes kept their heads low. The Federals had won the fight.[117]

The next day, Law prepared the division for another assault on the Federal entrenchments and awaited his orders from General Longstreet. Although they had suffered heavy casualties in the previous day's fighting, the men of Sheffield's (formerly Law's)

---

[116] Laine and Penny, *Struggle for the Round Tops*, pages 40, 67.

[117] Laine and Penny, *Struggle for the Round Tops,* pages 95-105; quote by John Dykes Taylor, page 19.

Brigade were in high spirits and ready to resume the fight. The fight, rather, would come to them. Meade thought that the Confederates who had been repulsed on the Union left were vulnerable and could possibly be destroyed. Thus, he ordered a cavalry attack on the Confederate lines. Brigadier General Judson Kilpatrick, a native of Gettysburg, formed his brigade of Federal cavalry south of Sheffield's line – outflanking the Alabamians.

Consisting mostly of Vermont horsemen, Kilpatrick's cavalry opened its attack in the open fields near the Meyers Farm southwest of the Bushman Woods. Skirmishers from the Forty-seventh Alabama immediately engaged the riders. Those skirmishers from Marshall County, Alabama, fell back toward the main line for reinforcements, and the pursuing Vermont horsemen followed into Sheffield's trap. Confederate sharpshooters, sent to assist the Forty-seventh Alabama, selected their galloping targets and dropped them with ease. The snipers bought Sheffield and Law enough time to form a battle line from the southern base of Big Round Top west to the Emmitsburg Road.

At 4:00 p.m., Kilpatrick's cavalry, both mounted and dismounted, assaulted the Alabamians' line in full force. The bulk of the fighting occurred in the Bushman Woods. The Alabama skirmish line held strong, allowing Law and Sheffield to pull the Alabama Brigade from the Big Round Top position back to the Emmitsburg Road near the Cullen House. The efforts of the Alabama skirmishers proved very significant. Lee had just begun the climactic assault on the Federal center, later known as "Pickett's

Charge". Had the Alabamians allowed the Federal cavalry to command their flank, Lee's entire army could have been enveloped.[118]

Many southern troops were left behind on the battlefield of Gettysburg when the army retreated on 4 July. John Newton Shirley, Third Sergeant of Company B., Forty-eighth Alabama, was wounded severely in the elbow during the assault on Little Round Top on 2 July. He lay on the battlefield for three days without food or water. A Gettysburg lady brought corn pone and clabbered milk to the wounded men in the vicinity of Little Round Top and Devil's Den, and Shirley recalled the meal as "the best meal I ever ate!" Later that evening he was moved to a surgeon's tent and his arm was amputated.[119]

Beginning on 5 July Lee's battered army retreated back to Virginia. Law's Brigade was marched to the Fredericksburg area and entrenched around the city's heights. There they prepared for a major counter-offensive by the Union Army.

---

[118] Laine and Penny, *Struggle for the Round Tops*, pages 136-142.

[119] *The Heritage of Marshall County, Alabama* (Clanton, Alabama: Heritage Publishing Consultants, Inc., 2000): pages 293-294.

Joshua Price

5

CAMPAIGNING IN TENNESSEE

"Remember boys, we are here to whip them!"
- General John Bell Hood, September 1863

After the Battle of Gettysburg, the Army of Northern Virginia and the Army of the Potomac were exhausted. Both armies halted further campaigning to recover from the extreme losses sustained during the three day battle in July.

The Forty-eighth Alabama, with the rest of Law's Brigade, marched to Fredericksburg, Virginia in early August 1863, and remained there until it was ordered to march to Richmond on 8 September. There, it boarded a train and headed south to reinforce Braxton Bragg's Army of Tennessee camped near Chattanooga, Tennessee.[120] While in Fredericksburg, the Confederate army continued its recuperation after Gettysburg. The wounded had time to heal and the rest of the men renewed training and drills. While off duty, many soldiers explored the town of Fredericksburg, only to find the town ravaged by war and inflation. "Everything is so high [priced] it is impossible to buy anything. A good mackinaw would be $300," wrote Captain Isom Small of Company E, Forty-eighth

---

[120] Hoole, *John Dykes Taylor*, pages 18-19.

Alabama.[121]

On 9 September the Forty-eighth Alabama, along with the Fourth and Eleventh Alabama regiments, reached Hanover Junction at 9:00 a.m., boarded the train for Richmond around 5:00 p.m. They camped near Manchester that night. There, they boarded a train and headed south to reinforce Braxton Bragg's Army of Tennessee camped near Chattanooga, Tennessee. On 10 September, the three regiments boarded different cars at Petersburg and reached Weldon at 3:30 p.m. After a five hour rest, the journey continued, and they reached Raleigh, North Carolina around 4:00 p.m. The men camped for the night. The regiments arrived in Charlotte at around 4:00 p.m. on 13 September, then continued to Columbia, South Carolina arriving around 4:30 p.m. The men immediately changed cars and proceeded to Augusta, Georgia, and arrived around 10:00 a.m. on 14 September. The train left three hours later, and the regiments arrived in Atlanta at dawn on 15 September.[122]

---

[121] Isom Burnett Small to wife, 8 August 1863. Letter reprinted in the *Gadsden Times* in April 1961 as an entry to its "Centennial Edition" celebrating the beginning of the American Civil War. Small was a graduate of the Presbyterian College in Oxford, Alabama and was trained as a minister. The *Gadsden Times* entry reported Captain Small "was carried dying from the battlefield of Chickamauga." Military records and his tombstone indicate he died at White Plains, Alabama on 24 June 1864 of disease. He is buried at Copeland Bridge Cemetery in DeKalb County, Alabama.

[122] Turner Vaughan, "The Diary of Turner Vaughan, Co. C., Fourth Alabama Regiment, C.S.A." *Alabama Historical Quarterly 18*, no. 4 (December 1956): pages 595-596.

The trip from Virginia was very strenuous for the Alabama troops. The cars in which they travelled were very cramped and uncomfortable. As the train made its way south, the people along the way waited with great anticipation for the arrival of the troops. The civilians prepared large picnics for the soldiers. This was very welcoming to the weary men, and they enjoyed the rest and relaxation with a hearty meal.[123] "Large crowds of ladies were assembled at different points, who greeted us with sweet smiles, kind words, and frequently with quantities of nice rations. The boys amused themselves with loud cheering as we flew past bevies of beautiful ladies who waved their snow white handkerchiefs to us," Captain Joab Goodson wrote.[124]

As the ragged soldiers stepped down from the trains when they arrived in Atlanta on 15 September, Colonel Sheffield observed the men and recommended they receive fresh supplies. Many of the men had not been issued new accessories since the regiment had arrived in Virginia in 1862, and most were without proper shoes and clothing. Sheffield had always sought the best for his regiment, or at least an equal share. While in Atlanta, the Forty-eighth Alabama, along with the rest of Law's Brigade, received the supplies it needed. The men received new shoes, hats, and other types of clothing. The freshly equipped men re-boarded their train and were shipped to

---

[123] Stocker, *From Huntsville to Appomattox*, page 133.

[124] William Stanley Hoole, ed., "The Letters of Captain Joab Goodson, 1862-1863." *The Alabama Review*, no.2 (April 1957): page 149.

Dalton, Georgia during the evening of 17 September.[125]

By the afternoon of 19 September, Law's Brigade formed their battle lines in the woods near Chickamauga Creek. As the men stood in battle formation, listening to the rattling of musketry and the booming of cannons only a thousand yards ahead of them, General John Bell Hood rode along the lines of the Alabamians. Since the brigade had not yet received its support batteries or its surgeons, Hood decided that it was not yet ready to go into battle. The regiment would wait until the next day. "We were in no condition to go into battle", wrote a member of the Fourth Alabama. The following morning, when the surgeons and artillerymen had arrived, the brigade formed a line of battle. Colonel Sheffield, mounted upon his "magnificent iron grey" rode to the front of the Fourth Alabama. In a very dramatic gesture Sheffield reeled around on his horse, held his hat high in the air, and shouted in that "tremulous and quavering manner of voice peculiar to him", "Forward, Fourth Alabamians, Forward! You have a name that will never, never die!" Law's Brigade entered the Battle of Chickamauga.[126]

Braxton Bragg, commanding the Confederate forces at Chickamauga, promoted Hood to command the Army of Tennessee .

---

[125] Hoole, *John Dykes Taylor*, page 20.
[126] Stocker, *From Huntsville to Appomattox*, pages 134-135. Sheffield managed to bring his horse, a gift from the men of the Forty-eighth Alabama, with him from Virginia. As the train moved south during the trip from Virginia, the horse managed to jump out of its cage and off the train. It was retrieved without any injuries.

Law was promoted to command Hood's division, and Sheffield took command of Law's brigade. Lieutenant Colonel "Mack" Hardwick of Company C was given command of the Forty-eighth Alabama. Hood made one last review of his Alabama brigade before he moved to Army Headquarters. "Remember boys, we are here to whip them!" he shouted as he rode along their lines.[127]

At 3:00 p.m. on 19 September, Sheffield formed his brigade's battle line south of the Brock Field. Its alignment, from south to north facing west, was the Fifteenth Alabama, the Forty-eighth Alabama, the Forty-seventh Alabama, the Fourth Alabama, and the Forty-fourth Alabama. Sheffield's (formerly Law's) brigade followed Bate's brigade as it advanced toward the Brotherton farm on the Lafayette Road.[128]

As the brigade approached the battlefield it came under heavy artillery fire. An exploding shell spooked Sheffield's "big iron grey" and it threw the colonel to the ground. Having injured his back, Sheffield was replaced by William C. Oates of the Fifteenth Alabama.[129] The battlefield at Chickamauga was, much as it is today, heavily wooded with occasional farmhouses and pastures. Roads made traveling the battlefield in 1863 easy for some troops. However, for most troops land navigation at the Battle of

---

[127] Laine and Penny, *Law's Alabama Brigade*, pages 144-145.

[128] Laine and Penny, *Law's Alabama Brigade*, page 147.

[129] Laine and Penny, *Law's Alabama Brigade*, page 148.

Chickamauga was difficult because some units marched through the green thick forests. By 3:30 p.m., there was much confusion in the ranks as the brigade pressed westward toward the Lafayette Road. Portions of the Oates' (formerly Sheffield's) brigade began mixing with and fighting alongside Tennessee Regiments. Nevertheless, the Confederates continued to push forward at a rapid pace.[130]

Oates' Brigade reached the Brotherton Farm along the Lafayette Road around 4:00 p.m. and met heavy Federal resistance along the ridge at the farm. They faced Ohioans and Kentuckians under the command of Colonel Charles G. Harker. After a fierce fight near the Brotherton Farm, Oates' brigade, along with Fulton's brigade, was forced to fall back across the Lafayette Road into the woods west of Brock Field. Harker's attack and defense in the Brotherton Field halted the Confederate advance, and Oates was forced to withdraw his brigade west, back to its former position south of Brock Field. It remained there until the following morning.[131]

During the night of 19-20 September, the armies on both sides received considerable reinforcements. Union troops from Chattanooga arrived and Confederate troops from Atlanta arrived. Colonel William F. Perry of the Forty-fourth Alabama received command of the brigade, and he prepared it for combat. This time,

---

[130] Laine and Penny, *Law's Alabama Brigade* page 150.

[131] Laine and Penny, *Law's Alabama Brigade.*, page 155; see also Stocker, *From Huntsville to Appomattox,* page 135.

with the support of artillery on either side, the brigade formed its battle line. From left (south) to right (north), the brigade stood as follows: the Fifteenth Alabama (Oates), the Forty-eighth Alabama (now under the command of Mack Hardwick), the Forty-seventh Alabama, the Fourth Alabama, and the Forty-fourth Alabama. By 6:30 a.m., Perry's (formerly Oates') Brigade was again ready for combat.

Hood's entire division did not move until approximately 11:00 a.m. when it was ordered forward to split the Union line at a gap near the Brotherton Farm. The brigade sent skirmishers from each regiment out front to protect its advance and to maintain contact with McNair's brigade – which was directly ahead of Perry's brigade on the Brotherton Road. Perry's brigade, at the front of Hood's Division, followed Robertson's Brigade. It moved west along the Brotherton Road and was then ordered to move northwest toward the Poe House. Here it engaged Indiana, Ohio, and Kentucky troops under the command of General John Brannan. Perry's regiments hit the Federals hard in the woods around the Poe House, and the defenders quickly retreated to the west. Perry followed the fleeing Federals, and Benning's Georgians closed in.

The skirmishing and fighting was very heavy in the Georgia woods. The thick wilderness filled with smoke, and sometimes caught fire, causing great navigational difficulties. Amidst the confusion, Oates' Fifteenth Alabama became disconnected from the rest of the brigade and fell in line with Robertson's Texas brigade south of the Dyer Road – Glenn-Kelly Road intersection. When Perry realized that Oates was no longer on the left flank, he wheeled the next regiment in line, the Forty-fourth Alabama, south to protect the brigade's left (southern) flank.

General Hood accompanied the Alabamians into the Dyer Field. Amidst a hail of bullets, he rode to the edge of a small patch of woods near Kershaw's brigade. A bullet found Hood's right leg and he fell from his horse. His thigh was broken by the ball. Hood was removed from the field and the leg was amputated. In less than three months the Texan had lost his left arm at Gettysburg, and now his right leg at Chickamauga.[132]

Perry's brigade continued to pursue the fleeing Federals, and it fought its way through the thick woods east of the Glenn-Kelly Road, emerging at Dyer Field. Dyer Field, a forty-plus acre open field which lay before the brigade, was guarded on the rises of its western edge by the guns of the Union XXI Corps' artillery. At noon, Perry's brigade, along with the brigades of Sugg and McNair, pressed their way into Dyer Field. As the brigade emerged from the woods at the Glenn-Kelly Road it immediately came under extreme artillery fire. Perry's brigade attacked the batteries near the tree line along the northern edge of Dyer Field. Protecting the batteries from Perry's brigade were the Twenty-sixth Ohio Infantry Regiment and

---

[132] Peter Cozzens, *This Terrible Sound: The Battle of Chickamauga* (Chicago: University of Illinois Press, 1992): pages 411-412. Hood never returned to the Army of Northern Virginia. He was given command of the Army of Tennessee and fought in the battles around Atlanta and Nashville.

the Fifty-eighth Indiana Infantry. The fighting between the Federal regiments and Perry's Brigade was swift and fierce.[133]

By 1:00 p.m. on 20 September, most of the Federal army was routed and was retreating north to the high ground called "Horseshoe Ridge." Most Confederate brigades concentrated in the woods east of the ridge and attacked it at approximately 1:15 p.m. Snodgrass Hill is the most northern of the spurs along Horseshoe Ridge, and its slopes made it a nearly impregnable position. The Confederate assaults lasted for over four hours, but the Federal army, commanded by George Thomas, held the ground. At dusk, the Confederate Army halted its attacks and readied for the next day of fighting. General Rosecrans ordered the remnants of the Federal Army to retreat from the battlefield, back to Chattanooga.[134]

George Thomas' defense of Snodgrass Hill had saved the Union army at Chickamauga from certain disaster. As Rosecrans' routed army retreated to Lookout Mountain after the Battle of Chickamauga, on 21 September, Longstreet prepared his Corps to follow-up the Confederate victory. Cautiously, Bragg ordered Longstreet to send a single division to harass the Federal rear.

---

[133] J. Gary Laine and Penny, *Law's Alabama Brigade*, pages 164-169; see also Cozzens, *This Terrible Sound: The Battle of Chickamauga*, pages 383, 397-416. Oates attached the Fifteenth Alabama to the rear of the Nineteenth Alabama of Dea's Brigade at The Tanyard south of Dyer Field. There, it engaged Lytle's Brigade at 11:45 a.m.

[134] James Longstreet, *From Manassas to Appomattox*, pages 456-457.

Longstreet pressed Bragg to attack Rosecrans with full-force, but Bragg was not convinced that Rosecrans was really retreating. He believed that the people of Chattanooga would be pleased to see the defenders of their city march triumphantly through the streets in grand review.

A Confederate soldier, who had escaped from Chattanooga after being captured during the fighting on 20 September, reported to Bragg the condition of the retreating Federal Army. Bragg inquired of him: "Do you even know what a retreat looks like?"

"I ought to know, General," replied the soldier, "I've been with you during your whole campaign."[135]

When Bragg ordered the Confederate Army forward, Chattanooga was surrounded and besieged. Longstreet argued that the failure to follow-up the victory at Chickamauga allowed the Federal Army in Eastern Tennessee to survive. The general was probably right. Bragg had given Rosecrans an opportunity to rest his army, prepare impregnable fortifications in and around Chattanooga, and receive reinforcements. The Confederates occupied the high grounds around Chattanooga, such as Missionary Ridge and Lookout Mountain, and laid the siege.[136]

Bragg's forces closed all major roads, railroads, and

---

[135] Stanley F. Horn, *The Army of Tennessee* (Norman: University of Oklahoma Press, 1952): page 277.

[136] Longstreet, *From Manassas to Appomattox*, page 462-463.

waterways entering Chattanooga. The Federal Army in Chattanooga had only one supply line remaining. The "Cracker Line", as it was called, originated twenty miles southeast of Chattanooga in northeast Alabama at the town of Bridgeport. From Bridgeport, supplies were sent north on the Tennessee River to Kelly's Ford. From Kelly's Ford, the line ran toward Lookout Valley to the town of Wauhatchie. From that point, supplies were carried north to Brown's Ferry, and then to Chattanooga.[137]

Longstreet immediately recognized the importance of closing the Cracker Line. On 8 October, he sent the Fourth Alabama of Law's Brigade to Wauhatchie. The men served as sharpshooters, targeting the mules of the Federal teamsters. On 9 October, Oates and the Fifteenth Alabama arrived at Wauhatchie. Rations in the valley were scarce. Many of the men, both North and South, were hungry and had not eaten in a couple of days. When the Fifteenth Alabama and a Federal regiment discovered an un-harvested cornfield near Wauhatchie, The two regiments of hungry soldiers exchanged shots as men attempted to gather the raw treat for their comrades. "I believe we got the most", recalled Oates. After a couple of days of foraging, Bragg sent the regiments of Law's brigade rations via pack-mules. Oates purchased sheep and cattle

---

[137] Peter Cozzens, *The Shipwreck of Their Hopes: The Battles for Chattanooga* (Chicago: University of Illinois Press, 1994): pages 54-57; for an interesting article on the "Cracker Line" please see Patricia L. Hudson, "The Old Anderson Road: Lifeline to Chattanooga." *Tennessee Historical Quarterly 42*, no. 2 (1983), pages 165-178.

from a local farmer, gained access to a cornfield, and impressed a corn mill. The brigade ate well for the remainder of the month.[138] The Forty-eighth Alabama, Forty-seventh Alabama, and the Forty-fourth Alabama regiments arrived on 15 October and remained near Wauhatchie until they were ordered back to the south side of Chattanooga to join the main body of the Confederate Army on 25 October.[139]

Hooker had moved north from Bridgeport to secure Brown's Ferry. The Federal troops arrived to fortify it on the morning of 27 October. Law responded and ordered his three regiments (the Forty-fourth, Forty-seventh, and Forty-eighth) to return to Wauhatchie to support the Fourth and Fifteenth regiments in case of an attack by Hooker. Longstreet ordered Brigadier General Micah Jenkins, Hood's replacement, to move his entire division to the west side of Lookout Mountain to await the probable attack by Hooker's army. The former commander of the Army of the Potomac arrived in the vicinity of Wauhatchie in the evening of 28 October.[140]

At 7:00 p.m., Sheffield ordered Captain Thomas Eubanks of Company D out front with pickets, and the Forty-eighth Alabama moved across Wauhatchie Creek. In moonlight, the regiment formed a one-rank line of battle in a nearby open field. The regiment then proceeded to the ridge in front of the field to guard the

---

[138] Oates, *The War*, pages 269-271.
[139] Hoole, *John Dykes Taylor*, page 22.
[140] Stocker, *From Huntsville to Appomattox*, pages 139-145.

road leading from Brown's Ferry to Kelly's Ferry. Eubanks' men found a large encampment of Federal soldiers nearby, and Sheffield ordered Lieutenant Colonel Mack Hardwick to bring up the regiment. McDuffey's Company engaged Federal pickets and captured eight prisoners.[141]

Eubanks and his pickets continued to press forward through the thick woods in the moonlight. Suddenly, they surprised four Federal pickets and the Confederates took their aim. The Federals, after realizing they were about to be shot, begged the Confederates not to shoot and offered their weapons. However, as Eubanks lowered his weapon and ordered his men not to fire, two of the "prisoners" suddenly turned their weapons and fired two shots into Eubanks' stomach. Watching their captain's fatal wounding, the Confederate pickets fired into the Federals – avenging the treacherous act.[142]

The Federals in the nearby encampments were alerted to heavy skirmishing occurring and desperately sought their attackers in the thick Tennessee woods. The soldiers were "anxiously peering through the darkness out of which all too soon the long dusky lines were described moving forward."[143] The skirmishing soon

---

[141] *OR, S1, V31, P1*, pages 228-230.

[142] *Guntersville Democrat*, June 1895.

[143] George W. Skinner, ed., *Pennsylvania at Chickamauga and Chattanooga: Ceremonies at the Dedication of the Monuments Erected by the Commonwealth of*

developed into intense fighting, and for three hours the battle "...roared up into a nocturnal holocaust verging upon the anguished tumult of hell."[144]

Sheffield was ordered to take command of the brigade and formed its battle line. From left to right, the men from Alabama stood in the following order: the Fifteenth Alabama, the Forty-fourth Alabama, the Fourth Alabama, the Forty-eighth Alabama, and the Forty-seventh Alabama. Sheffield took all precautions in the night fight. He ordered *vedettes* in front of each regiment's pickets, and one whole company from the Fifteenth Alabama stood guard on the right flank at the ridge. Sheffield then ordered all regiments to construct breastworks of logs and rails.[145]

Hooker's cavalry soon confronted Sheffield's defensive lines, but were quickly repulsed by numerous concentrated volleys. With little loss to Sheffield's men, the Federals fell back in great confusion, but rallied in a nearby wood. They made another charge, this time directed at the part of the line held by the Forty-eighth Alabama. The Federals were once again repulsed. Soon afterward, Law ordered the brigade to fall back to the field in which it had first formed its line of battle. After doing so, the brigade retired back

---

Pennsylvania (William Stanley Ray, State Printer of Pennsylvania, 1904): page 256.

[144] George K. Collins, *Memoirs of the One Hundred and Forty-ninth Regiment New York Volunteer Infantry* (Syracuse, New York: by the author, 1891): page 202.

[145] *OR, S1, V31, Part 1*, pages 228-230, Sheffield's report.

across the Wauhatchie Creek bridge.[146] After the engagement at Wauhatchie Creek, the Forty-eighth Alabama remained in that vicinity participating in picket duties and performing hard labor (likely fortifying).[147]

After the fighting ended and the brigade retired across Wauhatchie Creek, Sheffield hurried to the infirmary where Eubanks lay in agony. The following morning Eubanks sent his young wife, whom he had married only two weeks earlier while on furlough, a final message by Sheffield. The colonel stayed with his old friend, who expired later that morning. Sheffield had Eubanks' lifeless body wrapped in an oil-cloth and placed in an empty supply wagon. Sheffield drove the wagon, unescorted, home to Warrenton, Alabama where Eubanks' body was buried near his home.[148] Many of the men who served with Eubanks admired him. "He was a gallant and talented young soldier, and his memory has been always held in peculiar veneration by the people of his county."[149]

---

[146] *OR, S1, V31, Part 1*, pages 228-230, Sheffield's report; some sources report that Hooker's "cavalry" were not horses, but were, rather, mules. For details on this please see John K. Stevens, "Of Mules and Men," *South Carolina Historical Magazine 90*, no.4 (1989), pages 282-298.

[147] Hoole, *John Dykes Taylor*, page 22.

[148] *Guntersville Democrat*, June 1895; although the captain has a memorial tombstone in the Guntersville City Cemetery and also in Rome, Georgia, the location of his remains are in the original grave near his home in Warrenton.

Jenkins' Division was ordered to march toward Knoxville, Tennessee on 5 November to remove Federal forces under General Ambrose Burnside that were located there. The brigade boarded a train at Tyner's Station and moved north to Sweetwater, Tennessee. From Sweetwater the brigade marched to Loudon, Tennessee, where it arrived on 14 November. The following day it crossed the Tennessee River and marched to Lenoir's Station.[150]

Longstreet continued to pursue Burnside's forces on 15 November. Burnside halted his forces at Campbell's Station and awaited the approaching Confederates. Burnside's troops formed defense lines north of the Kingston Road – Concord Road intersection. Longstreet deployed his forces on the north and south sides of the Kingston Road.[151] McLaw's Division stood on the north side of the Kingston Road (Confederate left) with Bratton's command in the Confederate center. Anderson's and Law's Brigades, placed on the right of the Confederate line, were ordered to attack the Federal left. On 16 November, the Confederates proceeded with the attacks. Anderson's Brigade attacked first and the defenders retreated to a second line of defense. Law's brigade moved to the right of Anderson's and attacked the second Federal line. Law's attack on the second line of Burnside's defense at Campbell Station was not simultaneous with that of Anderson. It

---

[149] Hoole, *John Dykes Taylor*, page 22. Taylor called the death of Eubanks "treachery".

[150] Stocker, *From Huntsville to Appomattox*, page 148.

[151] Laine and Penny, *Law's Alabama Brigade*, page 207.

was not successful, and the Federals repulsed the Alabamians.[152]

Nevertheless, Burnside withdrew his forces from Campbell's Station and moved north toward the defenses prepared near Knoxville. Longstreet followed the Federal troops and surrounded them at a makeshift fortification called Fort Loudon. It was rumored among the men that Longstreet would try and bring Burnside to surrender rather than fight again. Burnside refused to surrender, however, and attacked. The following morning, 18 November, a column of Federal infantry marched under heavy fog directly in front of the Alabamians. The Confederate pickets began firing and soon Law's Brigade was firing into the Federal ranks. "We fired a deadly volley at close range, which was so unexpected that it caused a retreat to the cover of the Fort, leaving a number of dead and wounded." One of the wounded Federal soldiers captured by the Alabamians was the brother of a member of the Fourth Alabama.[153]

Jenkins blamed Law for the defeat at Campbell's Station. After the battle, Jenkins reported to Longstreet, "In a few minutes, greatly to my surprise, I received a message from General Law that in advancing his brigades he had obliqued so much to the left as to have gotten out of its line of attack. This careless and inexcusable movement lost us the few moments in which success from this point

---

[152] Laine and Penny, *Law's Alabama Brigade*, page 207.
[153] Stocker, *From Huntsville to Appomattox*, page 149-150.

could be attained."[154]

Longstreet sided with the favored Jenkins. As the hatred between Law and Longstreet intensified, the controversy was not kept quiet among the army. Officers, and even enlisted men, of the division were split over whom they wanted for the promotion to commanding officer. The commanding general made every attempt to court-martial Law for insubordination and refusal of orders.[155] Longstreet even contacted Confederate President Jefferson Davis and asked which of the two young generals should be promoted to Major General. Davis favored Law for the promotion because of his combat experience, but Longstreet favored Jenkins. Davis never formally promoted Law and left the situation to Longstreet. Jenkins retained his command.[156]

After hearing of this, Law tendered his resignation, which was denied. He was given a leave of absence to appeal to the War Department, and selected Oates and Benning as his defense counsel. All officers of Law's Brigade, with the exception of Colonel Perry, signed a petition to have the brigade removed from Longstreet's

---

[154] *OR, V31, P1*, page 526, Jenkins' report; The rivalry between Generals Law and Jenkins intensified after the Battle of Chickamauga in September. Law wanted to retain command of Hood's Division while the wounded general recovered from the leg wound suffered at Chickamauga. Longstreet, however, strongly supported Jenkins (who arrived after Chickamauga) to be division commander. Jenkins' commission was two months older than Law's and he assumed command of Hood's Division immediately after he arrived in September.

[155] Oates, *The War*, page 338.

[156] Freeman, *Lee's Lieutenants, Volume Three*, pages 309-314.

command and sent either to Mobile or back to Lee's Army of Northern Virginia. Longstreet was infuriated after reading the petition and ordered Law arrested.[157]

Davis understood the controversy was more than just a political distraction – the men were affected by it as well. He temporarily settled the Law-Longstreet-Jenkins debate by promoting Charles Field to Major General to command the division (effective 13 March 1864) and transferred the division back to Lee's Army of Northern Virginia.[158]

The feud between Longstreet and Law continued even after the First Corps returned to Virginia in May 1864. Longstreet placed Law under arrest and removed from his brigade once again. There was to be a hearing for the charges of insubordination that had been placed upon him months earlier, but Law claimed he was not aware of the charges and asked that he be provided with a copy of them by Longstreet's headquarters. Longstreet's aide refused to provide him with copies, and Law appealed once again to the War Department. He was released in time to participate in the battles around

---

[157] Oates, *The War*, page 338.

[158] Freeman, *Lee's Lieutenants, Volume Three*, pages 309-314; for a soldier's point of view on the officers' controversy please see Stocker, *From Huntsville to Appomattox*, page 149; for details of Field's promotion see Charles W. Field, "Campaign of 1864 and 1865," *Southern Historical Society Papers 14* (1886), page 542.

Spotsylvania Court House.[159]

---

[159] Oates, *The War*, pages 339-341.

6

GRANT VERSUS LEE

"...there will be a hot time in the Wilderness today, for there is blood in the sun."
- Robert Coles, 4th Alabama Infantry

President Lincoln appointed Ulysses S. Grant commander of all United States Army forces on 12 March 1864. Grant immediately went to Virginia to coordinate the 1864 campaign against Lee's Army of Northern Virginia. The campaign, called the "Overland Campaign," targeted the Confederate capitol of Richmond and its chief supply line from the city Petersburg. In April 1864, the Overland Campaign commenced and Grant moved toward Lee's defenders west of Fredericksburg. The engagement, at the site of the Battle of Chancellorsville a year earlier, became known as Battle of The Wilderness.[160]

Longstreet received orders from Lee on 11 April 1864 to return to Virginia with the divisions of the original First Corps of the Army of Northern Virginia. The divisions, those of Kershaw and Field (Hood), along with Alexander's artillery arrived in Charlottesville, Virginia by railroad on 14 April. The corps marched

---

[160] James McPherson, *Battle Cry of Freedom*, pages 718, 722.

from Charlottesville to Mechanicsville on 22 April and bivouacked until 2 May. Longstreet anticipated an attack on Gordonsville and ordered Field's Division to march north toward that town.[161]

Law's Brigade of Field's Division, commanded by Colonel William F. Perry (Forty-fourth Alabama), arrived in Cobham's Station, Virginia four miles south of Gordonsville, during the afternoon of 3 May.[162] The soldiers of Perry's (Law's) Brigade were in poor condition. Their numbers had not been replenished during the winter of 1863-64, and the men that remained were exhausted from their most recent transfer back to Virginia. Many of the men were hungry and had insufficient clothing. "A new pair of shoes or an overcoat was a luxury," recalled Law, "but the very fact that they remained with their colors through such privations and hardships was sufficient to prove that they would be dangerous foes to encounter upon the line of battle." Still, despite their commander's confidence, they were in no shape to fight.[163]

At 2:00 p.m. on 5 May, Perry's Brigade marched slowly up the Plank Road toward the fierce fighting raging near The Wilderness Tavern on the old Chancellorsville battlefield. The brigade halted near the Antioch Church at dusk, and bivouac, but it was still ten miles from the fighting. At 2:00 p.m. on 6 May the

---

[161] *OR, S1, V36, P1*, page 1054.

[162] Oates, *The War*, page 339.

[163] Evander M. Law, "From the Wilderness to Cold Harbor", in *Battles and Leaders of the Civil War, Volume IV*, page 120-121.

brigade resumed its march, at a brisk speed to arrive on the battlefield at sunrise. The brigade, following Kershaw's Division, cautiously approached in the thick dark woods. At daybreak, the battle resumed in The Wilderness but Longstreet's First Corps was still a few miles from the battlefield. The Corps' pace now increased to maximum speed.[164]

Litter bearers and ambulances packed the Orange Plank Road as they hurried the wounded from the battlefield. Perry's Brigade passed makeshift infirmaries. The grounds were littered with "depressing influences" of what lay in store for many of the men in the battle ahead.[165] Surgeons, busy amputating the arms and legs of unfortunate wounded soldiers, stacked the limbs unceremoniously beside the trees. Broken men lay strewn, moaning and crying, awaiting their turn under the saw. "It was perfectly appalling", recalled Bob Coles. Despite the horrible scenes, Perry's men pressed forward.[166]

As the brigade neared the battle line many men were wounded, and many others straggled behind. Close to the fighting, Robert Coles of the Fourth Alabama recalled, "…the sun, blood red

---

[164] William F. Perry, "Campaign of 1864 in Virginia," *Southern Historical Society Papers 7*, no.2 (1879): pages 50-51.

[165] Perry, "Campaign of 1864 in Virginia," page 51.

[166] Gordon C. Rhea, *The Battle of The Wilderness, May 5-6, 1864* (Baton Rouge: Louisiana State University Press, 1994): page 156; quote from Stocker, *From Huntsville to Appomattox*, page 160.

from the effect of the smoke of battle, was just appearing above the Wilderness…there will be a hot time in the Wilderness today, for there is blood in the sun."[167]

At 6:00 p.m. Perry formed his brigade into line of battle in a field near the Widow Tapp Farm, north of the Plank Road. The battle line was, from left (north) to right (south), the Fifteenth Alabama, the Forty-eighth Alabama, the Forty-fourth Alabama, the Forty-seventh Alabama, and the Fourth Alabama. Perry's Brigade formed behind a crude series of breastworks, brush and rotted logs. Simultaneously, Federal troops under Major General James Wadsworth's command trudged slowly through the thick woods north of the Plank Road, directly toward the Alabamians. Confederate scouts alerted Perry of these Federal movements on the left (north) flank of the brigade, and he immediately ordered Oates to detach his regiment and engage the threat. The Forty-eighth Alabama moved north of the breastworks at the edge of the wooded area near the Wilderness Run.[168]

The Alabamians marched east, down the sloping field toward the edge of Widow Tapp's Farm and into the Wilderness. By 8:00 a.m., the Forty-eighth Alabama, commanded by Major John Wiggonton of Company I, and the rest of Perry's Brigade swiftly engaged three of Wadsworth's Brigades – Rice's, Stone's, and Cutler's. "My front rank fired a volley without halting, and the

---

[167] Stocker, *From Huntsville to Appomattox*, page 160.

[168] Laine and Penny, *Law's Alabama Brigade*, pages 238-242.

whole line bounded forward with their characteristic yell. The enemy was taken by surprise. The suddenness of our appearance on the crest, the volley, the yell, and the impetuous advance caused them [Federals] to forget their guns. They returned only a scattering fire and immediately gave way."[169] Perry's men pursued the fleeing Federals. Wadsworth rode to the scene and rallied the One Hundred Fiftieth Pennsylvania Regiment, of Stone's Brigade, yelling "Give it to them, Bucktails!" The Bucktails sent one volley into the Forty-eighth Alabama and the Forty-fourth Alabama, halting the Confederate advance.[170]

Wadsworth's men rallied, and entered the fight once again, and the violent slugfest continued. The Forty-eighth Alabama, on the left of the line after the Fifteenth, moved north, and received concentrated fire. Its left was threatened by the One Hundred Forty-third Pennsylvania Regiment. "Many of the men were leaving the ranks and taking shelter behind the trees." The Forty-eighth rallied and held its ground until Oates returned. Perry recalled, "It was fortunate that the success of Colonel Oates had been so complete in his movement on the extreme left of the enemy…the Fifteenth regiment had arrived on the left at the crisis of the engagement, and delivered its decisive blow." That "decisive blow" came in one

---

[169] Perry, "Campaign of 1864 in Virginia," pages 52-53.

[170] Thomas Chamberlin, *History of the One Hundred and Fiftieth Regiment Pennsylvania Volunteers, Second Regiment, Bucktail Brigade* (Philadelphia: F. McManu & Company Printers, 1905): page 211.

murderous volley into the flank of the Bucktails, who immediately fled the field. The Alabamians recuperated as best they could and prepared for another assault on the breastworks.[171]

Among the casualties of the Forty-eighth Alabama was its Adjutant, Henry Figures. While leading the regiment in the fight against the One Hundred Forty-third Pennsylvania, the Huntsville, Alabama native was killed instantly when a bullet passed through his head. During the battle, his body was wrapped in his blanket and buried in a shallow grave beneath a peach tree near the spot where he fell. After the war, his family retrieved his remains and re-interred him near his home in Huntsville.[172]

Private Jesse Brackett, Co. B., Forty-eighth Alabama, was wounded severely in the right thigh by a Federal minie ball. "I lay on the battlefield three days and nights before I was removed by our litter corps." He remembered, "…crawling to a branch and the water there was almost totally red with blood from the wounded and dying soldiers". For the rest of his life he suffered from the wound, which caused him to limp in pain.[173]

---

[171] Laine and Penny, *Law's Alabama Brigade*, page 246; Perry, "Campaign of 1864 in Virginia," pages 54-55.

[172] Laine and Penny, *Law's Alabama Brigade,* page 253.

[173] Confederate Pension Records/Applications – 18 May 1914. The ball that struck Brackett was later found in his boot. He kept it as a painful reminder of his wartime experiences. He died in February 1916 and the ball is still in possession

Also among the Confederate casualties were Generals James Longstreet and Micah Jenkins. The two officers were shot accidentally (like Thomas Jackson and his staff a year earlier) by their own troops as they rode through the lines. Longstreet was severely wounded in the neck, and Jenkins was killed instantly.[174]

After the Battle of The Wilderness, Grant did not hesitate to move his army. The bulk of the Federal army moved southeast toward Spotsylvania Court House and prepared another assault, this time on Lee's right flank. For possibly the first time in the war, the pressure was on Lee rather than his opponent.[175]

Field's Division remained on the Wilderness battlefield until it was ordered to move south toward Spotsylvania Court House at 10:00 p.m. on 7 May. At dawn on 8 May, Field's Division arrived at the Old Block House at the crossroads of Shady Grove Church Road and Old Court House Road. Perry's Brigade was ordered to march north along the Old Court House Road to support Confederate forces already engaged with the Federals.[176] As Perry advanced north, he deployed to the west. The Fifteenth and Forty-eighth Alabama Regiments supported Henagan's left flank by engaging the Federals

---

of his decendents (2008). Being of Cherokee blood, Brackett was one of the few minorities within the regiment.

[174] Longstreet, *From Manassas to Appomattox*, page 564.

[175] Freeman, *Lee's Lieutenants, Volume IV*, pages 371-372.

[176] Laine and Penny, *Law's Alabama Brigade*, page 259.

troops under Brigadier General Lysander Cutler's command. The Alabamians took position behind log breastworks that were reinforced by dirt.

On 9 May, only minor skirmishing occurred between the two opposing armies as reinforcements arrived on both sides. At 10:00 a.m. on 10 May, Cutler ordered his men to charge the Confederate lines, a probe to determine its strength. The Federals poured from the woods in front of the Forty-eighth Alabama and the Fifteenth Alabama and were greeted by a murderous volley of Confederate musketry and canister. Among the killed was Brigadier General Rice, commander of Cutler's Second Brigade. Cutler's men fell back to the woods. The cheering Alabamians begged the shattered Federal brigades to attack once again.[177]

At 3:00 p.m., Cutler's men renewed the attack on the Confederate breastworks. Again, the Alabama regiments again poured volley after volley into the advancing Federals, and again, Cutler was violently repulsed by the entrenched Alabamians. "The blue columns pressed forward to the attack, and were sent back torn and bleeding, leaving the ground covered with their dead and wounded."[178] The ill-equipped Alabamians took advantage of the fresh supplies that lay now before them. When the Federals were withdrawn out of sight, many defenders leapt over the breastworks

---

[177] Laine and Penny, *Law's Alabama Brigade*, page 259; *OR, S1, V36, P1*, page 611, Cutler's report.

[178] E.M. Law, "From the Wilderness to Cold Harbor", *Battles and Leaders IV*, page 129.

and salvaged Federal rifles, ammunition, and other goods of use to them. The men rushed back over the walls and distributed their booty amongst the men. They dug in once again anticipating yet another Federal assault.

On the evening of 10 May, Perry's skirmishers were driven from the woods in front of the breastworks. A few moments later, Cutler's long blue columns marched out for another assault on the Confederate lines. The Federals advanced, under heavy fire, at their fastest pace. As they approached the Alabama regiments in breastworks, their lines "dissolved before the pitiless storm that met them." But they pressed forward over their dead and wounded comrades who lay heaped from the previous attacks. This time they engaged the Confederates in hand-to-hand combat, but once again, Cutler was forced to fall back.[179] The Alabamians cheered and taunted the repulsed Federals. The Confederates retrieved more supplies from the wounded Federals, and dug in for another assault.

Cutler's men emerged from the woods on 12 May. Under heavy artillery fire from nearby Confederate batteries, the assault broke before it was fully underway. The Alabamians were victorious.[180] After the fighting ended, a South Carolinian observed, "The [Federal] dead were literally piled in heaps and it is a wonder how anyone could have lived through those long hours of murderous

---

[179] Law, "From the Wilderness to Cold Harbor", page 129.

[180] Laine and Penny, *Law's Alabama Brigade*, page 264.

conflict in which trees were cut down and the breastworks themselves torn up by musket balls."[181]

Grant intended to take advantage of the superior number of troops wielded by the Army of the Potomac and force Lee to fight. Despite the staggering losses at Spotsylvania Court House, Grant did not hesitate to move his army. On 21 May, Grant marched his army southeast to Old Cold Harbor. As the armies moved, very little fighting occurred, with the exception of minor skirmishing. Lee realized he would have to win a game of speed to outmaneuver Grant.

On 31 May, Grant's army formed battle lines near Old Cold Harbor and attacked Lee's entrenchments there. Law's Brigade arrived on 2 June to reinforce Anderson's Georgians. The Alabamians hastily constructed breastworks overlooking the fields in which Grant's forces had attacked the previous two days.[182]

The Federals formed their lines in the woods opposite of the Alabamians. Remembering the slaughter of their comrades at Spotsylvania, the Union men wrote their names and addresses on small pieces of paper and pinned these to the backs of their uniforms. The men hoped these notes would identify them in case

---

[181] Gordon C. Rhea, *The Battles for Spotsylvania Court House, May 7-12, 1864*, page 304.

[182] Laine and Penny, *Law's Alabama Brigade*, page 269-272.

that soldier was killed in the coming attacks.[183] At 4:30 p.m. on 3 June, the Federals marched out of the woods, into the open marshy field, and pushed toward the Alabamians. The Federal brigades, under the command of General Martindale, marched in perfect order with bayonets fixed. When the Federals were seventy yards in front of the breastworks, Law's men took careful aim at their targets and poured volley after volley into the Federal lines. The fire was quick and deadly, and Federal soldiers dropped by the dozens. The Alabamians were in ranks four deep. As the front line fired, freshly loaded muskets were continuously handed to them. Law ordered the batteries to load canister and concentrate fire upon the Federals. The tight blue columns withered as "…heads, arms, legs, and body parts and muskets were seen flying through the air at every discharge." Martindale ordered the Federal troops to retire.[184]

Wholesale slaughter was nothing new to the Alabamians. In December 1862 they witnessed it at Fredericksburg. In July 1863, they assaulted the impregnable Little Round Top at Gettysburg. Since the fighting resumed in The Wilderness in May 1864, the Alabamians participated in the slaughter of thousands of Federal soldiers.

After the initial attack, Law ordered more ammunition

---

[183] McPherson, *Battle Cry of Freedom*, page 735.
[184] Laine and Penny, *Law's Alabama Brigade*, page 272-273; quote by Pinckney Bowles, "Battle of Cold Harbor", *Philadelphia Weekly Press*, 31 January 1885.

brought up from the supply trains. He understood that if the Alabamians fired as rapidly in a second assault their ammunition would soon be exhausted. When the Federals appeared in the fields again, Law's Alabamians reloaded empty muskets and took positions behind the breastworks. They discharged round after round into the oncoming Federal columns, and the Alabamians actually laughed. As soldiers dropped in the bloody field by the dozens, Law's men laughed as they continued the slaughter.[185]

The wounded Federal soldiers lay strewn amongst the dead in the open field before the Confederate breastworks. Cries for "water" rang from the wounded but no assistance was given to them. Any attempt made to help the suffering men was quickly ended by the accuracy of a nearby Federal sniper. It was rumored among the Confederates that Grant refused to send a flag of truce to collect the dead and wounded because he did not want to admit defeat. Sporadic showers followed the June 3 fighting. Then, intense June heat caused the dead bodies to decompose, filling the air with "a most sickening and nauseating stench." The wounded Federal soldiers lay where they fell until Grant asked Lee for a flag of truce on June 5. The Federals then collected their wounded and buried their dead.[186]

The battles between Lee and Grant in the Overland

---

[185] E.M. Law, "From The Wilderness to Cold Harbor", page 141.

[186] Stocker, *From Huntsville to Appomattox*, page 177.

Campaign were fought with heavy casualties. Continuous fighting from May 4 to June 12 resulted in the loss of over one hundred thousand men, sixty-five thousand Federal soldiers and thirty-five thousand Confederate soldiers. These staggering losses were a major part of Grant's grand strategy. Not only did Grant have Lee on the move, he forced the Confederate commander to wage a war of attrition, a war the Confederacy could not win. Grant quickly replenished his weakened lines, but Lee did not. Grant's hit-and-move strategy proceeded from The Wilderness to the outskirts of Richmond. Finally, Grant aimed his army toward Petersburg, and Lee entrenched the remainder of his Army of Northern Virginia along the eastern outskirts of the city.[187]

---

[187] Freeman, *Lee's Lieutenants, Volume IV*, pages 737, 743.

7

THE FINAL ROLL
"No men...ever fought more heroically."
- Colonel William C. Oates

Grant's forces cut the lines of communication north of Richmond as he moved toward Petersburg. Butler's Federal forces, no longer occupied with Beauregard's small force at Bermuda Hundred, cut the communication lines between Richmond and Petersburg. Lee understood that protecting the capitol was vital, and he moved his army to defend Petersburg.[188]

The Confederate First Corps left the defensive positions of Cold Harbor on June 13. On June 15, with minor engagements with the enemy on its march south, Law's Brigade entered the trenches east of Petersburg. Anticipating a major assault, the Alabamians labored continuously on the breastworks and trenches. Life in the trench was perhaps the worst the soldiers had experienced in their years in the Confederate Army. The June heat produced a drought. As the Alabamians stood in the trenches they padded the dirt into a fine dust; when the wind blew it created "suffocating" dust clouds. The trenches were filthy and full of potential diseases. Vermin infested the trenches increasing sickness. Rations brought to the

---

[188] Laine and Penny, *Law's Alabama Brigade*, pages 280-282.

men were usually rotted cornbread and rancid bacon. The poor attrition made the men sick – diarrhea and dysentery took a toll on them.[189]

In many areas around Petersburg the Federal lines were less than two hundred yards from the trenches. Sniping became a major part of the daily life of soldiers. Many soldiers, both Confederate and Federal, became casualties to sharpshooters who fired at any movement. In order to retrieve water and supplies from the rear, the Alabamians dug a series of zig-zag trenches to the rear to prevent a sniper from getting a clean shot at them. Immobility within the trenches and sporadic mortar fire had a heavy psychological impact upon both armies.[190]

Colonel William Perry commanded Law's Brigade as the latter recovered from a wound he received at Cold Harbor. The brigade stayed in the trenches until it was ordered to march north of the James River on 29 July. Field's Division marched through Petersburg at 1:00 a.m. on 29 July and at 8:00 a.m. Perry's Brigade boarded the Richmond and Petersburg Railroad. The men moved north to Rice Station, then the brigade crossed the James River and marched to New Market Heights, where it arrived near dusk. Its duty at this time was to engage any Federal troops moving on the railroad. The Federals attempted no demonstrations. The brigade

---

[189] Laine and Penny, *Law's Alabama Brigade*, pages 284-287; see also Freeman, *Volume IV*, page 531.

[190] Laine and Penny, *Law's Alabama Brigade*, pages 286-287.

made camp nearby on New Market Heights and stayed there until 13 August.[191]

While at New Market Heights, the Forty-eighth Alabama was assigned a new colonel, William Calvin Oates. James Sheffield had been absent on leave since the siege of Knoxville, and had resigned his commission. "Mack" Hardwick, commander of the Forty-eighth since it campaigned with Longstreet in East Tennessee, was captured by Federal forces at The Wilderness and was a prisoner. Oates now commanded both the Forty-eighth Alabama and the Fifteenth Alabama. The day he took command, Oates described the Forty-eighth: "…composed mainly of mountaineers, tough and tireless, and hence good material for soldiers, if properly handled. Its discipline was not very good, its equipment inferior, with every appearance of neglect." Oates immediately requisitioned all supplies necessary for the men, including clothing and other accoutrements. One day, Oates brought the Forty-eighth two wagon loads of fresh watermelons. This gift from their new commanding officer was well received by the hungry veterans.[192]

On 14 August, shots were exchanged between Federal forces and Field's Division. Those shots proved to be only preliminaries. On 16 August, Perry's Brigade, with the Forty-eighth Alabama and

---

[191] Laine and Penny, *Law's Alabama Brigade*, pages 287-288. George Washington encamped his troops at the same place after the British surrendered at Yorktown in 1781.

[192] Oates, *The War*, pages 372-373.

Fifteenth Alabama regiments leading, marched northeast to join the rest of Field's Division already engaged with Federal forces north of the Darbytown Road near Fussel's Mill. The Forty-eighth Alabama formed on the right of the Fifteenth Alabama in the woods south of the Darbytown Road. Federal artillerymen targeted the Alabamians as they arrived, and the shells splintered the trees above as the Confederates formed their battle line.[193]

Oates' two regiments marched toward the Federal lines and exchanged deadly volleys with them at a distance of less than sixty feet. The Fifteenth suffered severely from the volley, and Oates ordered the Forty-eighth to cross the fence and charge the left (south) flank of the Federal line. The charge was successful and the Federal line driven back two hundred yards.[194] Richmond reporters observed that the charge of the Forty-eighth Alabama was made "with an impetuosity that was almost irresistible."[195] Oates continued his advance with the two Alabama regiments. Upon reaching a trench, the Forty-eighth Alabama took positions and fired at the Federal soldiers. As Oates paced the line of the Forty-eighth, a bullet struck his right arm. "It struck me with such force that it turned me half around and stunned me…" Oates recalled, "…those large mini-balls strikes a hard blow!" Oates ordered Captain John W. Wiggonton of Company I to assume command of the regiment

---

[193] Laine and Penny, *Law's Alabama Brigade*, page 294.

[194] Oates, *The War*, pages 374-377.

[195] Richmond *Daily Enquirer*, August 31, 1864.

and charge the enemy lines. Oates, holding his mangled arm, leaned against a nearby apple tree and observed the fight. Wiggonton and the Forty-eighth pressed forward taking heavy casualties until it dispersed the enemy from the field. "No men…ever fought more heroically, on any field, than did the officers and men of that regiment near the Darbytown Road."[196]

Exhausted from the heat and loss of blood, Oates sat down beside the tree. Two men from the Fifteenth Alabama retrieved their colonel, without a stretcher, and took him to the infirmary. Oates was administered morphine and whiskey. The litter-bearers unrolled his oil blanket, placed him in it, and used it as a makeshift stretcher. At the infirmary, the doctors examined the wounded colonel. The surgeons administered chloroform and began to operate. Oates, aware of what was happening, recalled, "I knew what was being done when they sawed the bone; I heard it, but did not feel it." The following morning, Oates learned that he was not the only officer wounded. Wiggonton received a wound to the thigh during the charge. Unlike Oates, Wiggonton did not lose his limb, and returned to service.[197]

Oates, so impressed by the performance and ability of Captain Wiggonton, recommended him for promotion to major. Colonel Perry described that Wiggonton was "attentive to duty, has

---

[196] Oates, *The War*, pages 374-377.

[197] Oates, *The War*, pages 377-379.

the confidence of the command, and is of undoubted courage". Wiggonton was promoted to major and commanded the Forty-eighth Alabama for the remainder of the war.[198]

After the battle at Fussel's Mill Wiggonton and the Forty-eighth Alabama, along with the rest of Perry's Brigade, moved back to the New Market Heights where it entrenched.[199] The brigade spent the remainder of 1864 maneuvering and fighting a series of battles along the Darbytown and New Market Road. The first Battle of Darbytown Road occurred on 7 October 1864, the second on 13 October, and the third on 27 October. These battles involved the Confederates making fruitless assaults on heavily defended entrenchments. The heavy casualties weakened the already thin regimental ranks among the brigade. The Alabamians entered winter quarters north of Richmond on 27 October and did not see action again until April 1865.[200]

Grant's spring campaign opened on 1 April. His plan was to bombard Petersburg and move on Lee's right. At 9:00 p.m. Federal artillery shelled Petersburg, and Field's Division was hurried to meet Grant's infantry assault but it was too late. At 4:45 p.m. on 2 April Grant's infantry broke through the Confederate lines at Petersburg.

---

[198] Compilation of Military Records, Forty-eighth Alabama Infantry Regiment, C.S.A., eight rolls (microfilm), John Wiggonton records.

[199] Laine and Penny, *Law's Alabama Brigade*, page 298.

[200] Laine and Penny, *Law's Alabama Brigade* pages 304-313.

After nine long months of siege, Lee was forced to surrender both Petersburg, and ultimately, Richmond, to Grant.[201]

Longstreet's three divisions, including Field's, marched from Petersburg toward the Appomattox River throughout the night of 2-3 April. The Alabamians had little or no rations. The once proud soldiers that marched for countless miles under General Jackson in 1862 now were tired. Their morale sagged and they did not have the strength to fight as they once did. More and more men straggled on the march. Many of the companies in the Alabama regiments were reduced to less than one-third of their original strength. Three years of deaths, diseases, and desertions took a terrible toll on the regiment.[202]

Lee was surrounded by Federal infantry and cavalry. After six thousand Confederate soldiers were captured at Sayler's Creek on 6 April Lee declared, "My God! Has the army been dissolved?" Lee could muster only twenty-five thousand veterans to face Grant's army of over one hundred thousand.[203] Rumors circulated among the Alabama troops about Lee's intentions of surrender. As the inevitable moment approached, "every man in the regiment was so overcome with disappointment and grief, that they either fell down

---

[201] Laine and Penny, *Law's Alabama Brigade,* pages 326-327.

[202] Laine and Penny, *Law's Alabama Brigade,* pages 328-329.

[203] McPherson, *Battle Cry of Freedom*, pages 847-848.

or leaned against some support and wept."[204]

Lee surrendered his Army of Northern Virginia to Grant at Appomattox Court House, Virginia on 9 April 1865. Captain Alvin Oscar Dickson of Company A, Forty-eighth Alabama, refused to dismount from his horse after the surrender. After he was pulled off his horse by Union soldiers, he unbuckled his sword and threw it to the ground in disgust. The Brooksville, Alabama native marched the remainder of his company home in "perfect" military form, dispersing them upon arrival at Brooksville in May.[205]

For the Forty-eighth Alabama Infantry Regiment and the rest of Lee's Army of Northern Virginia, the war was over. The men simply went home. Most walked home. John Newton Shirley, of Company B, began his trek home immediately after the surrender. Along the way he met a man in charge of a warehouse. The man offered him all the sugar he could carry. Shirley took off his long johns, tied knots in the ankles, and filled them with sugar. He claimed the sugar gave him strength to walk home. Shirley's wife Eliza became anxious when she did not hear from her husband after the surrender at Appomattox Court House. She sought answers from a fortune teller who told her to be patient – for he would arrive soon riding a mule. A few days later, John Shirley arrived home on a

---

[204] Stocker, *From Huntsville to Appomattox*, page 192.
[205] *The Heritage of Blount County, Alabama* (Clanton, Alabama: Heritage Publishing Consultants, Inc., 1999): pages 175-176.

mule.[206]

James and John Garrard, two privates in Company C, encountered Union soldiers during their return home. The two brothers, spotted by Federal Infantry, ran into a nearby barn to hide. When ordered to come out and surrender they cracked and rubbed rotten eggs all over their bodies. When they opened the doors they begged the Union soldiers for help – telling them they had a disease that was eating away their skin. The potential captors were frightened and told them to stay away from them. The two brothers, after the Federals ran away, had a good laugh and walked on home.[207]

Grant's "Overland Campaign" of continuous warfare forced Lee to fight a war of attrition and also to fight on grounds chosen by the Federal commander. It can be argued that Grant "besieged" Lee's army on 4 May 1864 when the Battle of The Wilderness began. From May 1864 until the surrender in April 1865 Grant continuously attacked the Confederate defenders. Thus, Grant manipulated Lee and, to an extent, controlled the fate of Lee's army.

---

[206] Family legend.
[207] Family legend.

8
CONCLUSION

"...I hope He will still remember me and abundantly bestow His richest blessings upon me and hasten the time when we can meet in peace once more."
- *Pvt. John M. Anderson, Co. K*

After the war the survivors attempted to reassemble their lives as best they could. Most of the men stayed in northeast Alabama and the majority of them occupied themselves, according to census reviews, as farmers. Others, such as the Miller brothers of Company C, operated prosperous lumber and syrup mills. Many returned to little or nothing and moved west hoping to start new lives. A large number of men, including Samuel Boyd and Benjamin Kiker of Company B, moved to Oklahoma and Texas and raised their families. Numerous officers, including Sheffield, Sam Rayburn, Augustine Woodliff, and John Wiggonton, became public servants and pursued careers as politicians in both state and local governments.

Many soldiers did not return. Their widows, such as Elizabeth Anderson (widow of John Anderson of Company K), were left to fend for themselves and raise their families alone. Many moved west with surviving family members, like the widows of James and William Parrish, who fled their desolate farms in northeast Alabama. In the 1890's many widows filed for and received a very small state government pension. Although small in amount, oftentimes the pension served as the only source of income for the elderly women.

Colonel Sheffield, who deposited all his money in a Richmond bank in 1862, became financially crippled when the Confederate economy collapsed. After the war he again served as sheriff of Marshall County, one of his many pre-war positions, and also as a representative to Montgomery. It is interesting to observe Sheffield's life after the war. The former cooperationist, like most southerners, embraced the Lost Cause after the Confederate defeat. He became vocal against Federal occupation and Reconstruction. Despite the danger of being arrested by Federal troops, he frequently spoke to crowds against Reconstruction and the military occupation of the south. Before the war Sheffield opposed secession.[208]

On June 19, 1890, while working as a government clerk in Montgomery, Sheffield received an urgent message from his family in Guntersville to come home immediately. Arriving in Guntersville the following evening, friends alerted him that his invalid daughter's caretaker had sexually molested her. Sheffield, enraged, went to Warrenton to the home of Dr. May, the caretaker. After Sheffield confronted him, May admitted to the accusation. After a brief argument, Sheffield shot and killed May.

---

[208] The *Gadsden Times*, June 26, 1890.

May and Sheffield became acquainted during the doctor's brief service as medical examiner in the Forty-eighth Alabama. Sheffield trusted May to protect the interests of his family during his absence. May's actions not only physically violated the girl, but also Sheffield's honor and in some ways the camaraderie of serving in the Forty-eighth. Sheffield's violent response to May's actions reflects scholarly arguments made concerning honor and manhood in an honor-based society. Controversies between two men, despite their relationship, oftentimes led to a violent conclusion.

That evening Sheffield returned to Montgomery, sorted his affairs there, and returned to Guntersville to surrender. A brief trial, presided over by Judge Samuel Rayburn, occurred the following week. Rayburn, a former officer in the Forty-eighth Alabama, acquitted his former commander of all charges. Rayburn's loyalty to Sheffield is testimony to his respect for the colonel and his "oath" to honor. Sheffield returned to Montgomery where he worked the remainder of his life.[209] He died in Montgomery on July 2, 1892 and is buried in the Oakwood Cemetery.

Many of the Forty-eighth's survivors lived long and prosperous lives. George William Chumley, who scaled the ranks from private to lieutenant in Company B, lived ninety-eighth years. In 1866, while working as a mill hand in Mississippi, Chumley contracted a deathly case of malaria and was bedridden. After a few days he began reciting Bible verses and his fever broke – thus

---

[209] The *Gadsden Times*, July 10, 1890.

escaping death. Soon afterward he returned home to Collinsville where he preached for the next seventy-seven years.

On June 28, 1938 Chumley, with his attendant Carl Porter, boarded a Pullman train headed to Gettysburg to commemorate the seventy-fifth anniversary of the battle. Chumley, ninety-three years old at the time and the last known surviving member of the Forty-eighth Alabama, sat quietly in deep thought and rarely spoke – which was very much out-of-character for old veteran. Emotionally, he recalled his comrades and the sacrifices made during the heat of battle. "The older I get the more I get into the cause of the South," he told Porter. He acknowledged that the best never returned home.

After arriving in Gettysburg the nearly two thousand veterans toured the parts of the field at which they valiantly fought. Chumley immediately requested that Porter take him to the southern end of Seminary Ridge – the sight where Hood's Division launched its attack on Little Round Top and the Devil's Den. As the men drove across the fields, once were stained by the blood of young men of Law's Brigade, Chumley requested to stop. The white-bearded gentleman, dressed in his finest gray suit, hat, and walking cane, trekked slowly into the rocky valley at the base of Little Round Top. The area, made famous during the war by Alexander Gardner's photographs of dead Confederate soldiers, was called the "Slaughter Pen." To Chumley, it was the site where his regiment, led by Colonel James Sheffield, was fearfully punished as it attempted to turn the Union flank. The exhausted veteran sat upon a nearby rock and weeped. Porter, following a short distance behind, spoke but the old man did not answer. After a few moments the veteran's sobs ended. Porter noticed a blank gaze in the old man's eyes. To Chumley the sounds of the birds and festivities were replaced by screams, moans, and war. Porter patted the old man's back and Chumley asked for some water. Chumley recovered himself and the two boarded their car and drove to the top of Little Round Top.

Many veterans met on Little Round Top that afternoon. Chumley, needing a rest, noticed a Union veteran and a little girl sitting on a bench. Chumley asked if he could sit and soon rested with the two. The little girl, no more than five or six years old, looked at Chumley and said, "my grandfather is not mad at you, are you mad at him?" Both veterans laughed, and Chumley patted the little girl on the head and assured her that he was not mad at her grandfather. The two veterans sat for a long time talking. The Union soldier, whose identity is not known, served in a regiment that fought upon Little Round Top – possibly firing, at some point, into Chumley's regiment. Before parting ways, Chumley and his old foe swapped walking canes. For the rest of his life Chumley was always seen walking around town with his "Yankee Cane." Chumley enjoyed great health and stamina for the remainder of his visit to Gettysburg. On July 5 he began his journey home.

Reverend Chumley continued to preach and perform weddings for young couples. He enjoyed walking downtown, with his "Yankee Cane" of course, to greet people. When he saw kids playing war, he would tell them "war is not a game!" In December 1943 he contracted pneumonia. On Christmas 1943 the ninety-eight year old veteran succumbed to his illness. Many citizens in Collinsville still remember George Chumley – affectionately known as "Uncle Young." All remember him as being one of the kindest and most gentle people they have ever known.[210]

During its tenure the Forty-eighth Alabama Infantry Regiment, often overlooked by historians, developed from a very inexperienced group into a very distinguished combat unit. As a combat unit it is likely the Forty-eighth Alabama was not much different than any other Confederate infantry regiment. It was composed mostly of poor farmers and tradesmen. It received little training and learned the ways of war by experience. Like all regiments, it became accustomed to sacrifice, hardships, and death. After its first engagement at Cedar Run, officers from other regiments blamed the Forty-eighth for the collapse of the Confederate line, thus tarring the regiment with a bad reputation. After three years of combat in the Army of Northern Virginia, these men proved many times over their effectiveness as a combat regiment. The Forty-eighth Alabama fought in all of the infamous battles of the eastern theatre, save Chancellorsville. It can often be found at the most dramatic places on the battlefield. It assisted in the

---

[210] All information on George William Chumley was obtained through interviews and family papers collected over the years. Mr. Chumley's "Yankee Cane" fell into the possession of one of his great-granddaughters and is still a family heirloom.

repulse of the infamous Iron Brigade's attack against the unfinished railroad at Second Manassas. It was decimated near the Dunker Church in the West Woods at Sharpsburg. It took heavy casualties as it fought ferociously at Little Round Top. It held the center of the Confederate lines in the thick woods of Chickamauga Creek and The Wilderness. It mercilessly slaughtered Union troops at Cold Harbor and Petersburg in the summer of 1864. At its first engagement, the Forty-eighth mustered approximately 450 soldiers but could muster only 136 soldiers when it furled its colors at Appomattox Court House.[211]

---

[211] For Appomattox Statistics, see Laine and Penny, *Law's Alabama Brigade*, page 334.

Joshua Price

APPENDIX A
LETTERS AND PAPERS

JAMES LAWRENCE SHEFFIELD'S LETTER TO THE CITIZENS OF MARSHALL COUNTY, DECEMBER 1860

Mr. A.G. Henry,

My Dear Sir: Through you I address the people of Marshall County. The ordinance of secession has passed, as all of us have known from the first day of the convention that I would do. I opposed it from the beginning with all the skill and ability I possessed. In its present form it is less objectionable than we had reason to suppose that it would be, but still I voted against it to the last. It is done, however, and the only question which remains for us to decide is, whether we will take sides with our own State, or with the abolitionists. Right or wrong, the State has determined upon her policy. We cannot change that determination nor can we resist except by revolution and revolution is aid and comfort to the abolitionists. In such an emergency, I cannot doubt what will be the decision of the people of Marshall, but I do not choose to act until I hear from them. Our property, our wives, our children, our honor, are all at stake. We cannot abandon the State in the hour of its coming trials, without incurring the reproach of all posterity. We have lived under the shield of its laws in peaceful times, and it would be cowardly and unmanly to desert when the tempest is coming. Besides, desertion would not bring safety. The states around us have either gone out or the Union or will go, and if we attempt to resist our State Authorities, we will soon be in a position where we can look for no sympathy and expect no assistance. North Alabama will be converted into a slaughter pen, our property destroyed, and our children driven houseless into the fields. All this might be and would be submitted to by a brave people in a good cause, but when it is accomplished by the disgrace of desertion, the most reckless will pause before involving their county and themselves in such horrors.

We have not obtained what we wanted – what we thought we had a right to expect – but we must remember that this is a government of majorities and a majority has decided against us. We not only had the right, but it was our duty, to contend for our own opinion as long as opposition was available, but when a decision is made, it seems to me that there is no other honorable and manly course but to acquiesce in that decision, and resolve at all hazards and at every cost to sustain our State and repel every enemy who may assail her. This is my view, and upon that as an individual, I have a right to act, but as your representative, I am bound by your will, and therefore, I will not sign the ordinance until I hear from you. In my judgment it ought to be signed, and I hope you will so instruct me; but that is a matter for your decision, and to that decision I shall conform. My own conviction as to what ought to have been done, remains unchanged, but the decree has gone forth. Now let us stand by our arms, and let our motto be, 'United we stand, divided we fall'.

Your obedient servant,

Jas. L. Sheffield

REMARKS OF HON. J.L. SHEFFIELD IN THE STATE SENATE

On Thursday, Feb. 24 [1885], the Bill to Donate $5,000 to Aid in Completing the Monument to the Dead Confederate Soldiers of Alabama, Hon. J.L. Sheffield, Senator, said:

Mr. President:

Whilst I dislike to consume the time of the Senate, yet, sir, I feel it my duty to raise my voice on behalf of the bill now pending before this body for its action.

I am in favor of an appropriation of public funds to aid in the erection of a monument in memory of our gallant dead.

This honor is a duty we owe to those who have fallen, and will be a grand and patriotic tribute to the glorious cause in which they so nobly fell.

Mr. President – Go back to the traditionary shadows of the long forgotten past, before the pen of history had ever learned to paint the deeds of noble daring upon the immortal page, and the rude, but everlasting mound was erected by faithful and loving hands, to tell to future ages that unconquered warriors had died without a groan in some savage broil or had fallen unsurrendered on the field in defense of his wigwam home.

And, coming on apace, we reach the morning dawn of civilization, and while we find many material and humanizing changes, still the crude stone is erected in commemoration of fallen chivalry and departed manhood.

And as higher and higher rises the sun of civilization, chasing away with its Christianizing influences, the dark shadows of a wild superstition, and dissipating the hideous rites of savage worship and the revolting customs of barbaric government, we find rising higher and higher, and more polished and more imperishable, still the grand column, not so much to perpetuate the memory of departed greatness but to speak to unborn ages, in the lipless language of immortal gratitude, of that virtuous homage which the magnanimous living freely paid to the noble and undying dead.

Mr. President, there has never been an age or a people that did not gladly desire and proudly struggle to establish some imperishable monument to perpetuate, throughout the coming future, noble deeds of daring and bold fidelity to patriotism. Every land is ornamented by the polished shaft rearing its historic head amid the clouds of Heaven, thus connecting the material with the immaterial world, emblematic of the sacred tribute that the grateful living has paid to the immortal dead.

'Tis thus the past has ever established some permanent shrine at which future generations could pour out the sacred offering of virtuous sentiment.

Mr. President, shall Alabama be a land without a monument? Shall the glorious achievements of her past, live only in song and story? Shall the undying memory of her gallant sons, who placed their lives upon their country's altar, when the dark cloud of oppression hung like a pall around the citadel of our dearest rights, be transmitted to future ages, alone upon the page of history? Shall not we who are the living witnesses of our immortal past consecrate some spot upon the historic soil of Alabama, which will tell to the hearts of unborn millions that Alabama mourns with a proud sorrow over the graves of her faithful children? Shall we not erect an altar, even here, where amid the throes of a convulsed nation, was born the cause for which our sons and brothers died, to which distant ages may make their pious pilgrimage and learn amid the atmosphere a holy gratitude, the ever living lesson that oppression, defeat, and even death can never destroy or even dim the glory of a proud and bold manhood? Yes, a thousand times yes! Let the sun as he spreads his morning glory over the smiling vales and towering mountains of our cherished Alabama, look down upon at least one glorious shaft erected in grateful honor of the best blood that ever baptized the soil of any country.

Mr. President, if there ever was a land justified in the pride in her war tried and fallen braves, certainly Alabama will forever cherish the valor, will forever remember the courage, and will never forget the fidelity with which her noble sons stood by her in the darkest hour of her peril, their virtuous deeds of noble daring will ever hang like jewels around the neck of memory; their memories photographed by the meridian sunlight of a noble gratitude upon the tablets of our recollection, will ever hang in the sacred gallery of our hearts, where not even the dust of forgetfulness will ever be permitted to gather; these noble hands which carried aloft Alabama's standard unsullied by defeat and unstained by disgrace, upon a hundred bloody fields, though long since crossed upon a soldier's breast and cold in death will ever be warm in the eternal gratitude of a noble people.

Mr. President, when the eastern star of freedom grandly rose upon the darkness of oppression in 1861, it steadily ascended the high o'erarching Heavens till it rested above the political manger of Montgomery; and as it blazed with meteoric effulgence upon this historic city it invited with the enchantment of patriotism, our gallant sons and brothers from the mountains and valleys and prairies of Alabama. Did they come with giddy song or childless curiosity? No, but they come bringing with them the frank incense of unconquered manhood and the vigor of a grand patriotism to hall this newborn wonder, and with the unsheathed sword of freedom in their hands they recorded the solemn vow in Heaven that if the god of battles could be appeased by valor, or the tutelary saint of freedom could be moved by undaunted courage, the foul stain of an oppressor's track should never disgrace the soil of their beloved Alabama.

Mr. President, here upon this very hill, more than a quarter of a century ago, this august vow was uttered. Has it been kept? It has been written, signed, sealed, delivered and recorded, in letters of noble blood, upon every battlefield from Sumter to Appomattox, and has been filed away in the archives of immortality, where its manuscript will never grow dim by time, nor its parchment become discolored by age.

And, sir, if the angels of heaven are ever permitted to look down upon the joys or sorrows of earth, while they may have looked in pity upon this spot in 1861, yet they are still gazing with a magnetic pride upon the thousands of unmarked graves of noble Alabamians who fell in the defense of their country.

Oh, if there be within this earthly sphere a boon, an offering Heaven holds dear, 'tis the last libation liberty draws from the heart that bleeds and breaks in her cause!

Mr. President, I have seen somewhere these lines:

They were slain for us,

And their blood flowed out in a rain for us,

Red, rich, and pure on the plain for us.

And many years may go,

But our tears shall flow,

O'er the dead who died in vain for us.

Mr. President, will the Senate of Alabama refuse to vote the pittance asked in this bill, and that the electric wires will convey the news of Alabama refusal to appropriate money out of her State Treasury to aid in erecting a monument in honor of her gallant dead?

Then sir, this land of ours, whose very beauty is a derisive mockery, is the once famous sunny South – a land first immortalized by the valor of her sons and courage of her daughters; then, sir, it is a surrender of the only heritage that can or could be bestowed upon our gallant dead.

Mr. President, it may be that the sunset of age lends me mystical lore, but however this may be, I feel that I am nearing the evening of life. My lengthening shadow upon the sands of time warn me that my sun is rapidly sinking in the West; but oh, departed spirit of my youthful times, sound once more the grand slogan of gratitude and chase away the emasculating goblin of this age of ingratitude.

Mr. President, I have endeavored in my feeble way to discharge my duty to my dead comrades, and I will not only vote for the passage of the bill, but would to God that I could record that vote upon every hearth-stone and in every family Bible throughout the length and breadth of my unhappy country.

Senators, we are called upon to act in this matter, and in performing this action we should look alone to the pole star of duty to guide us. Let no idea of a time serving policy influence us, let not the siren charms of spurious policy swerve us, but let us, like the worthy fellow-citizens of our gallant dead, boldly perform the duties which the solemn obligations of the present hour enjoin upon us.

Mr. President, the grandest picture that the mortal imagination ever delineated upon the canvass of the world's admiration, is that of man standing high upon the rock of duty calmly looking down upon the surging billows of the world's ever changing opinions.

Let us not stop to ask if this is popular, if it is politic, if it is expedient, but let us ask the sole question, is it right, and to this answer let us respond, not with the hesitancy of doubt or with the alacrity of earnestness and with the firmness of courage.

Mr. President, it is said that man arrives at his conclusions by reason, and that woman arrives at hers by intuition, and the world's history proves the fact that woman is oftener right than wrong.

Mr. President – Man has given us traitors, from a Judas to a

Benedict Arnold, but the name of woman is unstained by treason.

> *Not she, with traitorous kiss her Saviour stung,*
>
> *Not she, denied him with unholy tongue.*
>
> *She, when Apostles shrank, could dangers brave,*
>
> *Last at the cross and earliest at the grave.*

    Yes, Mr. President, and behold the gallant efforts of our noble women in this holy enterprise. Men are halting, men are faltering, and doubting in this matter, but our noble women are struggling, bravely, on for the accomplishments of this grand enterprise. God bless them individually and collectively. If my arms were long enough I would gather every noble daughter of Alabama in one long, loving embrace.

    But, Mr. President, as this is a physical impossibility, and as I don't wish to be called extravagant, using the arms that nature has given me, I could freely embrace them all, one at a time, as they go on in their work of Heavenly gratitude.

                        James L. Sheffield

SELECT LETTERS OF RESIGNATION

Camp near Martinsburg

Sept. 21st, 1862

Sir,

I enlisted in the Confederate States service on the 8th day of April last. I have been in service ever since. I am fifty six years of age and have five sons who enlisted [in] the service. One sick and died. One was killed in Battle on the 31st [30th] augst. One wounded by slugs at same battle through both thighs and I myself a wound in my right thigh. My sons having familys, ours eight children, our two and four being men. I myself having been unfit for duty since the 9th day of Augst see it my duty and the intrest of the camp to tender my resignation as major of the 48th Ala. Vol. I therefore ask respectfully ask if be accepted.

I am respectfully, your most obd. svnt.

Enoch Alldredge

Major, 48th Regt Alabama

I have carefully examined Maj. Enoch Alldredge and find him inacceptable to perform the duties of an officer because of disease of the kidneys, debility.

Surgeon , 48th Ala Regt.

Joshua Price

Headquarters 48th Ala Vols

Camp near Rapidan, June 21, 1863

R.H. Chilton

A.A.&I. Gen'l

Sir, I have the honor to tender you this my resignation as Lt.Col. of the 48th Reg't Ala Vols for the following reason: I was wounded at the last Battle of Manassas on the 30th day of August 1862 by a gunshot wound through both thighs totally disabling me from the use of one of my legs and I have been able for duty for the last nine months and no probability of recovery so as to render me fit for field service. I therefore consider it my duty to resign and now respectfully ask it to be accepted.

I am very respectfully your ob'd. servant

J.J. Alldredge

Lt. Col. 48th Reg't Ala Vol's

July the 14th, 1862

Head Quarters 48 Regt Ala Vols

Camp Fullerson near Richmond Virginia

Hon Randolph, Secretary of War, C.S.

I herewith tender you my resignation as Captain of Co. A, 48 Regt, Ala Vols for the following reasons: 1st I am affected with Singlo Honea in such a form that it disables me so much. I find it impossible for me to undergo the dutys for which I am compelled to undergo having been in service three months and four days and have not been able for duty more than three weeks during the whole time. 2nd I have Direah in such a complicated form that medical aid seems to have no affect.

I am respectfully

Capt A.J. Alldredge

Commanding Co., A.

48 Regt, Ala Vols

I certify that I have carefully examined Capt. A.J. Alldredge and find him unable for camp duty because of chronic diarrhea and secure his resignation to be accepted.

July 14th, 1862

Joshua Price

Surg, 48th Regt Ala Vols

Headquarters, 3rd Brigade

Jackson's Army of the Valley

July 31st, 1862

Hon. Randolph, Sec. of War, C.S.A.

I have the honor to address you to inquire if my resignation as Captain commanding Co. A, 48th Regt, Ala Vols entered the 14th July Inst. has been accepted. If so be case forward it immediately as my state of health will not allow me to even keep up with the Regt.

Your obdt servt

Capt A.J. Alldredge

Camp Oct. 4th, 1862

Hon G.W. Rudolph

Sec of War

Sir

I hereby most respectfully tender my resignation as Lieut. Col. Of the 48th Regt Ala Vols for the following reasons. That I have an affection of the kidneys and rheumatism which renders me inacceptable of undergoing the duties of my office.

Most respectfully

Your Obdt Servant

A.A. Hughes

Lieut. Col. 48th Regt Ala Vols

Head Qurs. 48 Regt Ala Vols

Near Martinsburg Va.

Sept. 23rd, 1862

Hon Sec of War

Sir I have the honor herewith to tender my resignation as Capt of Compy G, 48th Regt Ala Vols. In justification of this resignation I have to state that it is in consiquince of extreme bad health caused from Dysphiksia and Chronic Diarrhea and renders me totally unfit for service and is in one case to such an extent that it threatens to terminate my existence.

I am sir verry respectfully

Your Obt Servt

A.L. Woodliff, Capt.

Comdg, Co. G., 48th Regt Ala Vols

I certify that I have carefully examined Captain Woodliff of Co G and find him incapable of performing the duties of an officer because of Dysphixia of lung and diarrhea.

Surg 48th Ala Regt

Camp 48 Regt Ala Vols

May 19, 1863

Sir,

I have the honor herewith to tender my resignation of the office of Capt. of Co. "E" 48th Ala Reg't unconditionally.

My reasons for so doing is that I am now under arrest charged with disobeying orders. In this that on or about the 10th inst. I was ordered to report with all men who had loaded guns to the Colonel of the reg't, that the same might be fired off. This I failed to do and the same evening one of the members of my company, in firing a cap as he supposed, fired the gun killing Private Wells of company "H". This gun was given to a private who had first returned to company from hospital, and who was informed that it was not loaded.

Very respectfully Your Ob't servant

F.M. Ross, Capt.

Co. "E", 48 Regt, Ala Vols

S. Cooper

Adjt & Inspr Gen'l

Head Quarters August 14th, 1862

48th Regt Ala Vols

Camp near Martinsburg, Va

Sir,

I have the honor herewith to tender my resignation as Captain of

Company G, 48th Regt. Ala Vols. The justification of this resignation I have to state that it is from physical inability caused from a wound received in my hip and groin by the explosion of a cannon, which naturally affected the spine. I am also severely affected with hemorage of the bowels. My spine being affected tenders me unfit for infantry service.

I am very respectfully your most obd servt

John S. Moragne

Capt Co. G, 48 Regt Ala Vols

Camp near Mechanicville

August 4th, 1862

To Hon Secty of War

Sir I herewith tender my resignation as Capt of Company H, 48th Regt Ala and respectfully ask its approval. The grounds upon which I ask an acceptance: I am fifty five years of age I have been in service nearly four months and find that my health is declining and that I likely will not permit of my remaining in the service which I very much regret.

I am very respectfully

Your obt serv

Reuben Ellis, Capt

Co. H, 48th Regt Ala Vols

SOLDIER'S LETTER HOME

Camp 48th Alabama Regt.

Near Fredericksburg, Va.,

August 8, 1863

Dear wife,

Your very affectionate letter of July 28th received, which gave me great satisfaction.

Well, dear, I do not know that I have anything of interest to write at present. I have a cough that troubles me a right smart. Cold has settled on my lungs. With the exception of this my health is good.

Everything is so high it is impossible to buy anything. A good mackinaw would be $300. Please try to send me some warm clothing before winter sets in.

I must cease. Take care of yourself and the dear children and do the best you can. Let us all put our trust in our dear Saviour who has promised to hear us and answer the prayers of all who call upon Him in prayer. Give my respects to all our neighbors and friends.

Joshua Price

I remain, your devoted husband until death,

I.B. Small, Capt.

OLD VETERAN WANTS NAMES OF COMRADES

The Democrat is in receipt of a letter from Isaac Newton Morris, a former citizen of this county, who now resides at Campbell, Texas, who wants to get in touch with some of his old comrades.

Mr. Morris served in the Confederate army and wants to locate some of his old comrades so he can make proof of his service.

He says he was born at Brooksville in 1842 and enlisted at Blountsville in Company A 48 Alabama under Col. Enoch Alldredge. He was wounded in the second battle of Manassas on August 30, 1862 and was not able to render further service after that date.

This old gentleman writes that he is almost blind and is disabled to earn a living and needs what ever help he can get from the state.

He will appreciate any information that will assist him in establishing his record of service.

Address him at Campbell, Texas

Houston Graves in same company and regiment.

This letter was written by PVT. Isaac Newton Morris, Co. A., 48th Alabama Infantry Regiment. It was located in the Blount County

Memorial Museum in Oneonta, AL.

NOTICE OF POSSIBLE DESERTION

Headquarters 48th Ala. Regt.

Feb. 18th, 1865

Hon. J.C. Breckinridge

Sec. of War

Sir,

I respectfully ask that ordnance Sergeant John D. Taylor of the 48th Regiment Alabama Vol. be dropped from the rolls as Ordnance Sergeant for prolonged and continued absence without leave. He received furlough of indulgences for thirty (30) days on the 18th of November 1863, to go to Marshall County Alabama. He has not returned or been heard from since; and from all the information that I have, I do not think he intends to return to the regiment. I therefore ask that he be dropped from the rolls and his name be sent to the Enrolling Officers of Marshall County Alabama.

I am sir very respectfully your obedient servant,

J.W. Wiggonton

Joshua Price

Maj., Commanding Regiment

Camp near Suffolk, Va.

April 21st, 1863

Mrs. Foster –

I wrote you a few lines about ten days ago which I hope you received, in which I stated I would soon write you again. I commenced to write to you yesterday but was so interrupted that I burned it. We were under furious fire yesterday and for near ten days past. On day before yesterday the Yankees attacked a fort of ours and took it with two companies of the 44th Ala. On yesterday evening Gen'l Hood called for a hundred men from each brigade as volunteers to retake the fort – making three hundred men – they were warned of the hazardous undertaking. The men volunteered and were placed under command of a Capt. from this brigade. I first volunteered to take command of the men from this regt. I then offered my services at Brigade Headquarters where they were accepted – I was then honored with the command of all the volunteers from this brigade, being right of the line. We moved off about dark to make a night attack on the fort but; low and behold, when we got near it we met Gen'l Hood who to the mortification of many of us, informed Capt. Couzins that he had postponed the attack until tonight. We awoke this morning and found the Yankees had evacuated the fort.

To tell you the truth my "Good Friend", I would have been proud of the priviledge of dying on such a field engaged as I would have been, in command of one hundred men, who volunteered to die. In one night we would have been victorious and gloriously so. I will tell you that on Monday the 13th first I had the honor to head one of Gen'l Hood's expeditions consisting of three companies of this regt. The order was to advance our companies until we drew fire of the enemy. There were two Capts. and one Major present. I placed myself twenty paces in front of our line to command foreward, the men and officers followed, the enemy immediately opened with round shot, shell, grape, and canister, but still we advanced. I led until every officer refused to go further for the reason that we had fulfilled the order. We were very near the Yankee fortification and had the men and officers followed I should have led them to where they could have picked off Yankees with muskets. I then asked for three men to volunteer and we went to the creek in front of their works and took a few fires and went back to our commands. I have a desire to be in one more heavy engagement and if it should please God for me to survive it, I think I may have reason to be more thankful to God and all my friends who have given their prayers for my safety. Oh! how ungrateful to God we all are that while our friends and comrades in arms have fallen and we are spared to move to new victories. We do not seem to appreciate the splendid task before us. My heart is in my community now and my thoughts to my

Joshua Price

dear brother whom I learn is dead. "God giveth and God taketh away" – Blessed be the name of the Lord. My health is good and my spirits encouraged with this letter. You will convey my best and heartfelt wishes to your family and from my hands to accept the same yourself.

Your friend,

T.J. Eubanks

This letter was written by a member of the Alldredge family many years ago. A copy of it was sent to me circa 2003 by an Alldredge family lady – I do not know her name either. I have done my best to translate the poor handwriting with as much accuracy as I can:

Enoch Alldredge

"Enoch Alldredge, oldest son of Andrew and Leah Chaney Alldredge was born May 16, 1807 in Bledsoe County, Tenn. and came to Alabama with his parents in December 1816. He grew to manhood here and married Amelia Pace, born June 24, 1811 and died March 18, 1864 at their home in Brooksville, Ala. Amelia was the daughter of John and Zipparah Kerby Pace of Kentucky.

In 1836 Enoch Alldredge was a private in Captain Musgrove's Company.

In 1862 he raised a company and at the organization of Forty-eighth Alabama Infantry was elected Major of it. Enoch's second son, Andrew Jackson Alldredge, secured the enlistment of enough men to be elected Captain of Company A of the 48th Regiment, Alabama Volunteers, in the service of the Confederate States for three years or during the war unless sooner discharged. It is a fact that Enoch Alldredge had 7 sons in the Confederate Army and <u>each one</u> volunteered to serve in the <u>Forty-eighth company</u>. Son-in-law Bill Suttles did not. John Pace Alldredge volunteered in Co. A – Auburn, Ala. Jesse J. volunteered in Co. K. 19th Ala. Infantry – Huntsville.

Hiram Wright Alldredge volunteered in Co. B. 19th Regiment. There is no other record on earth concerning him aside from my grandparents' Bible.

If you have used Memory A. Lester Alldredge's <u>Aldridge - Bracken History</u> you noted Garland Smith Alldredge listed as a son of Col. Enoch. Hog Wash! He was the son of Jacob Chaney Alldredge, Enoch's brother. I knew Garland as "Shug". He married Louisa Scott of Brooksville. I attended his <u>funeral</u>.

Van Buren Alldredge, son of Enoch, enlisted at Warrenton, Marshal County, Ala. Enlisted April 7, 1862, and served with company A – 48th Ala. Infantry. He was with John Pace near Richmond when the latter died after a two week's illness Aug. 10, 1862. He sent word to Cedar Run to his father Enoch to take John P's body back to Alabama but Enoch and son Jesse J. had been wounded. Van Buren had John Pace buried in a country churchyard.

An old Baptist (Primitive) preacher conducted the services at 4 o'clock. A sad event – Van Buren was killed in the Second Battle of Manassas or Bull Run, Virginia Aug. 31, 1862.

Let me add – Hiram Wright was taken prisoner at the Battle of Missionary Ridge 1863. He was taken to Fort Delaware where he died. It is not known when.

Alabama Dept. of Archives and History has no record on Jarvus? Patrick Alldredge. It is a cinch that he was in the 48th co. He returned to his family in Blount Co.

The youngest of Enoch's sons in service was Taylor Winfield nicknamed "Ned". He was born in 1845 and volunteered as soon as he could. He came out unscathed. He died in Cullman Co. Ala. And was buried at **xxx** Chappell.

Enoch and son Jesse J. returned to Alabama after resigning official duties at Cedar Run, Virginia (a fact) to recuperate. He was promoted to Lieut. Col. Over five senior Captains for meritorious service on that field. Being wounded so badly he was crippled the remainder of his life. He was appointed recruiting officer for North Alabama which position he occupied until the close of the war. Enoch was also involved with recruiting men with Jesse J.

Enoch Alldredge was called Col. Enoch Alldredge during the war and to the present time. I don't know enough about military rules to tell you why. I read how Andrew J. earned his title of Captain.

Bruce, we can't get Enoch elected to the Hall of Fame because there is no picture of him. I know from the size of his vest he was a small man. He was clerk of Salem Primitive Baptist Church at Brooksville, Ala. From 1834 to 1879. I do know he rated high in his church.

He would rise to the occasion when his integrity was challenged and invited a man into the ring to settle it. I can see the same spirit in myself. What I have written you concerning my grandfather David Alldredge's father and brothers is the truth.

It is common knowledge in Alldredge families that Enoch Alldredge, his seven sons and one son-in-law fought in the Forty-eighth Co. Bill Suttles did not."

Joshua Price

PENSION AFFIDAVIT OF
SGT., THOMAS JEFFERSON PARRISH, CO. C

(Typed Verbatim)

The State of Alabama, Franklin County

Before me, S.J. Petree, Judge of Probate in and for said County, came personally Thomas J. Parrish who, being by me first duly and legally sworn, says, on oath, that he volunteered for service in the Confederate Army, November 1861, was mustered in February, 1862 Company "C" 48th Ala. at Guntersville, he an three brothers, all in the same company, and he served in the same company to the end of the war. He further says that he did not desert the Confederacy, but served to the end and was discharged at Appomattox C.H. He further says that his brother Isaac Parrish was killed at Cedar Mountain, Va., June or July 1862, that his brother William was killed at the second battle of Manassas, that his brother James was killed at the battle of Sharpsburg, Pa. He further says that he was with all three of his brothers when they were killed, and that during the whole of the time, from start to finish he did not come home, and was only abscent from his company one time, and then had a sore foot, and got permission from his Captain, and was abscent only ten days. He further says that he was wounded in the hip in a picked fight in West Va. and received a scalp wound at Manassas, but did not go to the Hospital either time, that he was in the Seven days fight, Manassas, Sharpsburg, Fredericksburg, Stroudsburg, Cullpepper and others.

(Thomas Jefferson Parrish) sworn to and subscribed before me, this 21st day of October, 1914.

SJ Petree

Judge of Probate

When he referred to James' death, he meant Gettysburg not Sharpsburg. James was wounded on 2 JULY 63 at Gettysburg and died 26 JULY 63 as a prisoner at Fort Delaware in New York City.

Joshua Price

APPENDIX B

SELECT BIOGRAPHIES

Colonel James Lawrence Sheffield

Nicholas Sheffield, a native of Virginia, came to Huntsville, Alabama in 1818 and was employed as a carriage-maker. Mary Martin Sheffield, a homemaker, migrated to Huntsville about the same time from her home in North Carolina. James Lawrence Sheffield was born to Nicholas and Mary on December 5, 1819 in Huntsville. James was raised in this town and educated in the schools of Madison County. At the age of eighteen James left his parents' home in Huntsville and relocated in the Claysville area of Marshall County, Alabama, where he became employed for the next four years as a clerk in a local merchants' store.

Sheffield adopted the political ideologies of the local Democratic Party and became vocal in his beliefs at a very young age. He was elected Sheriff of Marshall County at age twenty-five in 1844. He held this post until he was elected to represent his county in the state legislature in 1855. During the presidential campaign of 1860 Sheffield campaigned vigorously for Stephen A. Douglas and the Democratic Party. During this election he constantly toured the counties of Northern Alabama giving very influential speeches in favor of Douglas.

After Alabama called for a secession convention in 1861, Sheffield, along with Arthur Campbell Beard, were elected to represent Marshall County. Sheffield, a member of a group called "Cooperationists" did not want to secede from the Union. This group wanted to resolve the issues with the Federal government by peaceful means, not by war. Beard immediately voted in favor of secession. Sheffield, however, left his vote to the people that he represented in a letter written from Montgomery on January 15, 1861.

The letter sent to Marshall County represented the ideas of most of the people in the south during the years of secession. Sheffield used this eloquent letter to appeal to people of all social statuses in Marshall County. Following his influence, the people Sheffield represented from Marshall County immediately replied in favor of secession. Sheffield, the obedient public servant, signed the order of secession without. He immediately became a staunch secessionist. He became a front-runner in secession politics and states' rights in Northeast Alabama.

Sheffield returned to Guntersville immediately following the convention to join a close friend of his, Thomas J. Eubanks, as officers in the newly formed Ninth Alabama Volunteer Infantry Regiment. He remained in this regiment until April 1862.

During the period from 1844 to 1861 Sheffield owned a very profitable plantation in Marshall County that was operated by Negro slaves. This business allowed him to build enormous wealth and prestige. In the spring of 1862 Sheffield used both of these attributes, and his friend Thomas Eubanks, to form and finance a new regiment of Confederate volunteers in Marshall County.

Eubanks owned and edited Marshall and Blount County's local newspaper - "The Marshall Eagle". Sheffield authorized Eubanks to promote in this newspaper the raising of new companies consisting of local men. Outfitting this new regiment of volunteers would cost Sheffield $57,000 of his own money – a testimony to his new wealth. When the volunteers arrived at Warrenton on April 6, 1862, Sheffield was unanimously elected colonel of the regiment. They deployed to the aid of Robert E. Lee in Northern Virginia.

Sheffield participated in many engagements over the next eighteen months as colonel of the Forty-eighth Alabama Infantry Regiment. He was wounded severely in the leg during the opening engagement at Cedar Run, Virginia and was absent recovering during the Second Manassas fight and did not return to the regiment until the day after the Sharpsburg fight. He commanded the regiment from that point without absence until he was severely affected by artillery concussion during the Knoxville Campaign – forcing his retirement from the military.

Colonel Sheffield reached his militaristic pinnacle during the Gettysburg and Chickamauga Campaigns. During those two fights he commanded Law's Brigade after Major General John Bell Hood was wounded. Many men noted his bravery and valor in their reports and diaries. He was as inspirational in combat as he was in politics. All those who served with him revered him.

After the war ended he returned to Marshall County to try and reassemble his life and accept the defeat of the south as best he could. When Federal troops and politicians occupied the south Sheffield became enraged. He traveled across northern Alabama frequently holding meetings speaking to crowds about the atrocities of Reconstruction. He fully embraced the Lost Cause and was recognized as one of the few former confederates that would speak in behalf of the Democratic Party despite the dangers of being arrested by the Republican occupiers.

He was elected to the State Senate in the 1880's. After his debacle with Dr. May (as described in the text) he returned to Montgomery and lived the rest of his life. He died July 2, 1892 in Montgomery in is buried in Montgomery's Oakwood Cemetery.

Major Enoch Alldredge – Company A.

Enoch Alldredge was born in central Tennessee May 16, 1807. In 1816, shortly after Jackson's war with the Creek Indians, he traveled south with his family and settled in the Brooksville Community in western Blount County, Alabama. He attended local schools at an early age developed into a very intelligent young man.

In 1828, at the age of twenty-one, he was elected Justice of the Peace for Blount County. He held this position until 1836 when he joined Colonel Musgrove's company of volunteers as a private, serving in the Seminole Wars in Florida. When he returned home from that war, the same year, he was elected to the Alabama State Legislature. He served twenty-six sessions in this position.[212]

When secession came to Alabama in 1861 Alldredge quickly went to work forming a company to fight for the Confederacy. He was the grandson of a Revolutionary War soldier wounded at the Battle of Cowpens in 1781. If soldiering was not second nature to this peaceful country gentleman, being a leader certainly was. "Enoch Alldredge was a self-made, close observer of men, possessing much sagacity and energy". He used his political influence and soldiering nature to form a company of soldiers from the Brooksville community. He took this company, which included four of his sons and a son-in-law, to the Marshall County Court House in Warrenton on April 7, 1862 and formed Company A of the Forty-eighth Alabama Volunteer Infantry Regiment. He was elected Major of the regiment.

Alldredge was seriously wounded in the calf at the regiment's first engagement at Cedar Run, Virginia. The wound partially crippled him. After he returned from the war in October 1862, Alldredge tried to resume his life in Brooksville. He used his

---

[212] Dictionary of Alabama Biography.

political skills to promote recruiting in northeast Alabama. When his son Jesse was seriously wounded and returned home, the two recruited soldiers together and promoted the Confederate effort until the war ended in April 1865.

Alldredge managed the family farm in Brooksville until he was elected to the State Legislature again in 1874. During his public service career he was elected to various House Committees including the Committee on Public Education. He was also appointed Chairman of the House Committee of Accounts Statisticians. He was held in high regard by all that he served with both in both the military and government. It is likely that he embraced the Lost Cause as well. No one has yet surpassed Alldredge's time served in the legislation. Enoch Alldredge died at his home in Brooksville on November 22, 1879 and is buried in the family cemetery near his home.[213]

Major John William Wiggonton – Company I

John William Wiggonton was born May 28, 1828 in Carroll County, Georgia. His father, Colonel Isaac Wiggonton, served in the

---

[213] Fowler, Hugh Alvin. "Enoch Alldredge: Statesman". A letter written by an anonymous family member – seems to be a sort of eulogy to Enoch Alldredge. The author is unknown but it is obvious that she knew Enoch well. A copy of the letter is in possession of the author.

War of 1812. Colonel Wiggonton settled his family in present-day Cleburne County, Alabama sometime between 1840-1850. Although little is known about Wiggonton's childhood it can be determined that he and his six siblings were reared under a strict father. This trait was reflected in both his military and public careers.

Wiggonton enlisted in the "Newman Pounds Guards" in 1862 – they later formed Company I of the Forty-eighth Alabama. The name "Newman Pounds" was derived from the surnames of Captain Wiggonton's in-laws – Newman and Pounds. He was elected as its first commander (Major) upon its creation April 26, 1862. In late 1863 Colonel William Perry (commanding Law's Brigade) cited Captain Wiggonton as "unquestionably the senior Captain of the Regiment…he is attentive to duty, has the confidence of the Command, and is of undoubted courage." This recommendation rewarded Wiggonton with the promotion to major and also sole command of the Forty-eighth Alabama. He held this post until the close of the war.

Major Wiggonton was wounded twice during the war – first at Sharpsburg, Maryland on September 17, 1862 and again at New Market in August 1864. Perhaps the most painful wounds he received were the loss of two brothers. Green Henderson Wiggonton was killed at Gettysburg on July 2, 1863 and his brother Matthew, wounded severely at Sharpsburg, was captured on November 28, 1863.

Wiggonton was well respected by Colonel William Oates. Oates, commanding both the Forty-eighth and the Fifteenth Alabama regiments during the fall of 1864, promoted Wiggonton to the rank of Major to assure he would get command of the Forty-eighth. He served with distinction as one of the regiment's best soldiers, and remained in command of the regiment and was responsible for its surrender at Appomattox Court House.

During the war Wiggonton reprimanded the regimental ordnance sergeant, John Dykes Taylor, for failing to return from furlough on time. He ordered Taylor relieved of his post. He believed in discipline – a trait inherited from both his military training and from his father.

After the war Wiggonton returned home to Cleburne County and he immediately entered public service. He served as Judge of Probate in that county for many years. He was well-respected and was a member of the Lebanon Congregational Methodist Church. Major Wiggonton died at his home on September 20, 1891 and his wife followed five years later. He is interred in the adjoining cemetery of his church.[214]

---

[214] **"The Heritage of Cleburne County, Alabama"**. **Article written by Wandagene Jones. Photo Courtesy of author of article; Oates,** *The War Between the Union and the Confederacy.*

Captain Alvin Oscar Dickson – Company A

Alvin Oscar Dickson was born in Blount County on May 4, 1839. He married Missouri Alabama Alldredge, daughter of Enoch Alldredge, on April 7, 1860. He joined on April 7, 1862, alongside his father-in-law and brothers-in-law, in the company raised by Alldredge that became Company A of the Forty-eighth Alabama Volunteer Infantry Regiment at Warrenton. He served, like all soldiers who volunteered in 1862, a term of three years to end of war.

Dickson quickly rose from the rank of private to Captain. At Gettysburg he commanded the pickets who covered the assault of Law's Brigade upon Little Round Top on July 2. Although he was wounded at Chickamauga, he was one of the few men who served the duration of the war without missing any major engagements. At the surrender of the army at Appomattox Court House, Union soldiers forcibly removed him from his horse. Once he was dismounted he threw his sword to the ground in disgust. Captain Dickson was responsible for surrendering the remnants of Company A of the Forty-eighth Alabama at Appomattox Court House – a muster totaling less than twenty men. He marched the remainder of the men home "in perfect military style" and dismissed them upon arriving home.

After the war, Dickson returned home to Brooksville where he resumed his career in farming. He operated a dry-goods store in Brooksville and became one of the town's postmasters. His signature appears in many of the pension affidavits applied for by Confederate soldiers – or their widows - after the war. He was a very influential member of his community and very supportive of the men and the families of the men that he served alongside during the war. It is evident from his writing that he fully embraced the Lost Cause.

In September 1913 Captain Dickson received a letter from a gentleman by the name of Thomas Owen in Montgomery requesting a written statement of the experiences of the Forty-eighth Alabama Infantry Regiment. Captain Dickson, ever so dutiful, obliged the gentleman. In October he wrote a detailed account of his experiences in the war. As he wrote, memories came back to him. "Oh how lonely and sad to call back to memory those days" he wrote. His reflections of the regiment offer some of the best accounts of the regiment that have managed to survive these many years.

Captain Dickson died at his home in Brooksville in 1923. He is buried, along with other members of the Forty-eighth Alabama, at Salem Church in Brooksville, Alabama.[215]

---

[215] "The Heritage of Blount County, Alabama." Article written by Lottie Painter Hudson; Letter from Alvin Dickson to Thomas Owen of the Alabama Department of Archives and History. The original letter is in the Forty-eighth Alabama's file at the ADAH.

Sergeant John Dykes Taylor – Company D

John Dykes Taylor enlisted as a private in Company E, Forty-eighth Alabama Infantry at Warrenton on April 7, 1862. Due to his experience in the warehousing and dry goods business during his civilian career, and perhaps some political influence, he was appointed regimental ordnance sergeant. He served at this position until late-1864 when he returned and never went back. Before his death in 1888 he compiled a series of letters that were altogether a narration of the regiment's combat experiences from its conception in Guntersville to the bloody Grant versus Lee campaign around Fredericksburg, Virginia in May 1864. This series of letters, first published by the Montgomery Advertiser after his death, became the official unit history of the Forty-eighth Alabama. Although his letters were edited and re-published by the University of Alabama in 1985, no official unit history has been produced offering historians "new" information on the regiment – until now.

Taylor was born in Georgia on May 8, 1830. He was the son of a veteran of the War of 1812 and the grandson of a Revolutionary War veteran. His youth was traditional to the times – he grew up farming and was educated at home.

He settled in Jackson County, Alabama in 1850 and worked as a store clerk. He studied law while in Jackson County and was admitted to the bar in 1857. Apparently not very good at his new trade of practicing law, he was employed once again in the wholesaling business in 1860 after moving to Guntersville. He worked at this position until he joined the Forty-eighth Alabama at Warrenton in the spring of 1862.

Taylor was thirty-two years old, had a wife, and son less than a month old when he enlisted in the Confederate Army in April 1862. He returned home from the Confederate Army in December of 1863 and a dispute sparked between himself and Major John William Wiggonton. Wiggonton dismissed Taylor as a deserter. After the war he resumed his former position as a wholesale clerk. In 1869 he was elected Justice of the Peace and Notary – a position he held until his death. He erected a large warehouse in Guntersville in 1885 (apparently near Gunter's Landing) and sold various types of produce there for the remainder of his life.

He died at his home in Guntersville on May 9, 1888, the day after his fifty-eighth birthday. His obituary described him as "a good citizen, kind neighbor, affectionate husband and father, and an honest man" and was "admired by those who knew him best."[216]

Joseph R. Hughes – Company B

Joseph R. Hughes was born March 14, 1842 in Gadsden, Alabama. His family was one of the earliest families to settle in the Gadsden area. He was, in fact, the first male child born in Gadsden.

---

[216] "The Heritage of Marshall County, Alabama". Article submitted by Judy Taylor Reed; Larry Joe Smith, editor. *Guntersville Remembered*. Creative Printers, Inc., Albertville, Alabama: Creative Printers, Inc., 1989, page 32.

Hughes lived on his father's farm near present-day Attalla until he was fifteen years old.

In April 1862 he made the trip from Gadsden to Guntersville to enlist in Company B, Forty-eighth Alabama Infantry Regiment. He was wounded twice in his tenure with the Forty-eighth: once at Second Manassas and again less than three weeks later at Sharpsburg. Hughes participated in all engagements with the Forty-eighth Alabama until his discharge from the army in October 1862. Although in a different regiment he returned to the Confederate Army in November and finished out the war with that unit. Hughes' new unit was engaged in the western theater of the war – fighting in battles in Mississippi, Tennessee, and Georgia. Hughes saw combat action in both major theaters of the war. He surrendered in North Carolina in May 1865 with General Joseph Johnston's army.

After returned to Gadsden after the war he engaged himself in quite a few business endeavors and politics. He worked as a clerk in a dry goods store, was in the milling business, real estate, hotels, and natural resources. Hughes built the elaborate Exchange Hotel in Gadsden and also the first steam operated flourmill in Gadsden. He was elected Clerk of the Court in 1874 and 1880. He was also a member of the Gadsden City Council through the 1890's.

He was a well-respected man in Etowah County throughout and was very influential within the community. He was a member of the First Methodist Church and also the Emma Ransom Camp of the United Confederate Veterans. Hughes died October 14, 1921 at his home in Gadsden at the age of seventy-nine.[217]

---

[217] "Confederate Veteran", 1921, page 431; *The Heritage of Etowah County, Alabama.*

APPENDIX C
COMPANY ROSTERS

Company A - Raised in Blount County, Alabama and mustered at Warrenton Court House in Marshall County, Alabama on April 7, 1862.

1 - Adair, Andrew P. – enlisted as a private and ranked out as a corporal; was wounded in the wrist at Gettysburg.

2 - Albright, A.C. – enlisted as a private.

3 - Albright, Jermiah – enlisted as a private; dismissed in April 1863 from Chimborazo Hospital.

4 - Alfred, Calvin – enlisted as a private

5 - Alldredge, Andrew Jackson – son of Major Enoch Alldredge; selected as a Captain by the men at time of formation of the company.

6 - Alldredge, Enoch – elected Major at time of formation; resigned in October 1862 due to wounds received at Cedar Run on August 9, 1862.

7 - Alldredge, Jesse J. – son of Major Enoch Alldredge; elected First Lieutenant at time of formation; would rise to Lieutenant Colonel before his resignation in June 1863; was seriously wounded through both thighs at Second Manassas; after the war he and his father would return to Blount County and promote the war effort as an army recruiter.

8 - Alldredge, John Pace – son of Major Enoch Alldredge; enlisted as a private and died of disease in Richmond August 19, 1862.

9 - Alldredge, Nathan – enlisted as a private and discharged due to ill health in 1863.

10 - Alldredge, Van Buren – son of Major Enoch Alldredge; enlisted as a private and was killed atSecond Manassas.

11 - Allred, S.B. – enlisted as a sergeant.

12 - Bailey, Richard – enlisted as a private; died of disease on July 14, 1862; never experienced combat; buried in Virginia.

13 - Bailey, Zachariah – enlisted as a private

14 - Balers, C. – enlisted as a private; mortally wounded at Chickamauga and died of his wounds at home in Blountsville on October 10, 1863.

15 - Barnard, Patton Abner – elected as First Lieutenant at formation; forced to resign January 29, 1863 because of rheumatism.

16 - Beam, Hiram – enlisted as a private; wounded at Second Manassas.

17 - Beam, Whitfield – enlisted as a private; paroled at Appomattox Court House.

18 - Booker, S.M. – enlisted as a private; wounded at Cedar Run; died of disease at a Lynchburg, Va. hospital October 21, 1862.

19 - Brasseale, J.H. – was selected to be First Sergeant at time of formation but was reduced in rank to a private due to desertion; dropped from the army in June 1864.

20 - Burgess, A.J. – enlisted as a private; wounded seriously at Cedar Run and again May 31, 1864.

21 - Burgess, J.S. – enlisted as a private and promoted to sergeant January 30, 1863; wounded in the right arm at Chickamauga and

paroled at Appomattox Court House.

22 - Burgess, William C. – enlisted as a private; after Cedar Run, he complained of diarrhea; he would be killed at Second Manassas on August 29, 1862.

23 - Cain, R.A. – enlisted as a private; paroled at Appomattox Court House.

24 - Cargo, Albert – enlisted as a private and died at Auburn July 12, 1862.

25 - Chandler, H.J. – enlisted as a private and promoted to sergeant.

26 - Christopher, J.E. – enlisted as a private; captured at Deep Bottom.

27 - Conner, Henderson S. – enlisted as a private; killed at Gettysburg.

28 - Cook, J.B. – enlisted as a private; spent 1862 and 1863 sick at the Chimborazo Hospital suffering from dearrhea; no evidence exists that he ever returned to the regiment.

29 - Cook, J.J. – enlisted as a private; transferred from Chimborazo Hospital to Atlanta and there died October 10, 1863.

30 - Cook, William – enlisted as a private; died of pneumonia March 11, 1863.

31 - Cooper, C.M. – enlisted as Quartermaster Sergeant; discharged from service April 19, 1863.

32 - Cornelius, Bailey – enlisted as a private; absent without leave December 31, 1862; died in hospital of typhoid fever October 26, 1863.

33 - Curtis, C. – enlisted as a corporal at formation.

34 - Davis, S.W. – enlisted as a private; captured after being wounded in the arm at Sharpsburg resulting in its' subsequent amputation at a Union hospital.

35 - Densmore, A.Y. – enlisted as a private; discharged in 1862.

36 - Densmore, Samuel P. – enlisted as a private and promoted to sergeant; deserted on December 10, 1862; returned soon after and remained the rest of the war; paroled at Appomattox Court House.

37 - Dickson, Alvin Oscar – enlisted as a private and promoted to Captain; paroled at Appomattox Court House.

38 - Doty, Abraham H. – enlisted as a private; captured at Gettysburg; arrived at Lookout Point, Maryland a prisoner of war "under an assumed name" – an attempt to escape if exchanged?

39 - Drake, Elisha F. – enlisted as a private; absent sick for the majority of the war; captured at Macon, Georgia either April 20 or 21, 1865 and paroled there.

40 - Folkes, J.D. – enlisted as a private; died of pneumonia September 17, 1862 on the Mississippi River on his way home from prison camp.

41 - Fortner, E.F. – enlisted as a private; deserted in 1863.

42 - Fortner, Levi – enlisted as a private; wounded at Cedar Run and again at Manassas; deserted with his brother in 1863.

43 - Goldsmith, William – enlisted as a private; detailed to hospital work for remainder of war 1862.

44 - Goodsly, Jesse – enlisted as a private.

45 - Graves, Charles – enlisted as a private; paroled at Appomattox Court House.

46 - Graves, Cunningham – enlisted as a private and promoted to sergeant; paroled at Appomattox Court House.

47 - Graves, Franklin – enlisted as a private; wounded May 31, 1864.

48 - Graves, Houston – enlisted as a private; wounded at Second Manassas and The Wilderness; captured at New Market; exchanged March 11, 1865.

49 - Graves, Huston – enlisted as a private; deserted March 1863.

50 - Graves, Randolph – elected Second Lieutenant at formation of company; eventually promoted Captain.

51 - Green, J.P. – enlisted as a private; wounded at Cedar Run.

52 - Grigsby, F.G. – enlisted as a private; paroled at Appomattox Court House.

53 - Grigsby, J.A. – enlisted as a private.

54 - Gunter, James – enlisted as a private; wounded at Second Manassas.

55 - Gunter, Kincheon – enlisted as a private; wounded at Second Manassas.

56 - Heranton, C.T. – enlisted as a private; deserted regiment July 18, 1862.

57 - Hicks, B.F. – enlisted as a private; absent sick 1863-64.

58 - Holley, J. – enlisted as a private.

59 - Hood, Edward Columbus – enlisted as a private; deserted

Joshua Price

December 10, 1862; returned; promoted to First Sergeant during the Battle of Chickamauga; wounded at The Wilderness; absent without leave after July 1864 (likely killed).

60 - Hopkins, S. – enlisted as a private; sick at a hospital in Petersburg, Virginia May 7, 1863 complaining of rheumatism; returned to duty May 21, 1863.

61 - Huddleson, J.N. – enlisted as a private; deserted December 10, 1862.

62 - Hyatt, Abraham – enlisted as a private; transferred from Hunt's Artillery Battery to the regiment on December 27, 1864; paroled at Appomattox Court House.

63 - Jenkins, Lewis M. – enlisted as a private; suffered from typhoid fever in Chimborazo Hospital, Richmond October 19, 1862; died of pneumonia on June 12, 1863.

64 - Jones, J.W. – elected Lieutenant at formation of company and promoted to Captain; captured while absent without leave near Huntsville, Alabama October 13, 1863; released after taking the oath of allegiance in Sandusky, Ohio June 21, 1865.

65 - Kelsoe, J. – enlisted as a private.

66 - Kelsoe, Thomas – enlisted as a private; suffered from typhoid fever and was placed in a hospital in July 1862; would spend the rest of his service time sick.

67 - Kirby, Dolphus – enlisted as a corporal and reduced in rank to a private May 20, 1863; absent without leave after October 1863; took the oath at Nashville, Tennessee October 22, 1864.

68 - King, James N. –enlisted as a private; killed at Chickamauga.

69 - King, John – enlisted as a private; captured at Champion, Mississippi May 16, 1863; sent to Fort Delaware in New York City March 22, 1864.

70 - Kitchens, J.T. – enlisted as a sergeant; transferred to Company G, Twenty-third Georgia Infantry December 31, 1862.

71 - Latham, E.H. – enlisted as a private; absent sick December 15, 1862; returned to the regiment; paroled at Appomattox Court House.

72 - Latham, J.M. – enlisted as a private.

73 - Latham, John – enlisted as a private.

74 - Latham, Uriah Y. – enlisted as a private; killed at Cedar Run.

75 - Lipscomb, John P. – enlisted as a private.

76 - Maloney, J.H. – enlisted as a private; sick at hospital in October 1862; present the rest of the war.

77 - Mason, T.P. – enlisted as a private; absent sick during most of the spring and summer of 1864; deserted from the regiment July 27, 1864; taken as a "rebel deserter" to Beverly, West Virginia where he took the oath of allegiance April 14, 1865.

78 - McCaghren, P.J.B. – elected Third Lieutenant at formation of the company; promoted to First Lieutenant October 29, 1863; dropped from rank November 4, 1863; wounded severely at The Wilderness; paroled at Appomattox Court House.

79 - McCollum, J.N. – enlisted as a private; absent sick during 1863 and 1864.

80 - McDonald, Aaron – enlisted as a private; died of disease in October 1862.

81 - McFreeman, John – enlisted as a private.

82 - Mitchell, J.L. – enlisted as a private; absent sick with diarrhea

Joshua Price

after December 1863.

83 - Mitchell, John – enlisted as a private; wounded at Cedar Run; paroled at Appomattox Court House.

84 - Moore, Matthew M. – enlisted as a private; died of typhoid fever July 25, 1862.

85 - Morris, J.N. – enlisted as a private; wounded at Second Manassas; absent without leave after February 1863.

86 - Nation, David – elected First Lieutenant at formation of company; resigned July 10, 1862 because of age and hernias; never left Camp Auburn, Alabama.

87 - Nation, W.J. – enlisted as a private; captured in Eastern Tennessee January 10, 1864; sent to the Rock Island Barracks; transferred to Camp Chase, Ohio in February 1864; exchanged; died in a Richmond hospital from chronic diarrhea March 5, 1865 – one day after he was admitted.

88 - Neel, Wesley – enlisted as a private; killed in action at Chickamauga September 20, 1863.

89 - Neely, David – enlisted as a private; died from fever at Camp Auburn, Alabama June 1, 1862.

90 - Neely, William – enlisted as a private; wounded at Second Manassas; absent without leave during 1863 campaigns; returned in 1864 and was a part of the regiment until his parole at Appomattox Court House.

91 - Partin, George W. – enlisted as a private; wounded at Sharpsburg; died of nephritis February 1, 1863.

92 - Partin, J.W. – enlisted as a private; sick at Winchester, Virginia hospital August 9, 1862; wounded and captured at Gettysburg; exchanged; captured August 14, 1864 at New Market.

93 - Partin, T.M. – enlisted as a private; detailed to work at a shoe factory in Richmond; took oath of allegiance in Washington, D.C. January 6, 1865.

94 - Partin, W.H. – enlisted as a sergeant.

95 - Presley, William – enlisted as a private; absent without leave after July 1863.

96 - Pullen, Hampton – enlisted as a private; absent without leave after July 1863.

97 - Ratliff, J.M. – enlisted as a private; promoted to Second Sergeant July 31, 1863.

98 - Roden, J.S. – enlisted as a sergeant and demoted to private; wounded at Second Manassas; killed at Chickamauga.

99 - Rogers, A.T. – enlisted as a private; sick in hospital suffering from lung disease after June 1863; dropped from the rosters September 23, 1864.

100 - Shelton, W.H. – enlisted as a private; died of typhoid fever in a Richmond hospital March 20, 1863.

101 - Spears, Thomas – enlisted as a private; wounded at Cedar Run; absent sick in a hospital from October 1862 to January 1863 suffering from chronic diarrhea; captured near Knoxville on Novmeber 30, 1863; died from bronchitis as a prisoner of war January 4, 1864.

102 - Stepp, William – enlisted as a private; present through most of the major campaigns; paroled at Appomattox Court House.

103 - Stewart, Milton – enlisted as a corporal and promoted to sergeant; paroled at Appomattox Court House.

104 - Stewart, William G. – enlisted as a corporal; died of chronic diarrhea June 20, 1864.

105 - Stroud, Emanuel – enlisted as a private; deserted April 21, 1864.

106 - Taylor, A.F. – enlisted as a private at age fifty; discharged suffering from chronic rheumatism October 9, 1862.

107 - Taylor, T.C. – enlisted as a private; killed at Second Manassas.

108 - Thomas, D.N. – enlisted as a private; died of pneumonia in Richmond July 10, 1863.

109 - Thomas, D.A. – enlisted as a private.

110 - Thomas, J.P. – enlisted as a private; wounded at Chickamauga and died from wounds in Marietta, Georgia December 3, 1863.

111 - Thomas, Nathan – enlisted as a private; absent without leave after February 1863; returned to the regiment September 1, 1863; wounded at Chickamauga; wounded severely at The Wilderness.

112 - Thornsbury, A. – enlisted as a private; died of typhoid fever October 1, 1862.

113 - Thrasher, Thomas – enlisted as a private; sick in hospital most of 1862; died of fever December 10, 1862; note: he was present for duty during October 1862.

114 - Tidwell, Clayton C. – enlisted as a private and promoted to corporal; taken as a "rebel deserter" by the Army of the Cumberland; given the oath of allegiance in late September 1864, and ordered to "remain north of the Ohio River".

115 - Tidwell, S.R. – elected First Lieutenant at formation of company; resigned due to a fractured tibia June 10, 1863.

116 - Upton, William B. – enlisted as a private; captured near

Knoxville, Tennessee January 3, 1864; sent to Rock Island Barracks January 17, 1864; took oath of allegiance in Chattanooga, Tennessee March 3, 1865.

117 - Van Horn, A.J. – enlisted as a corporal; wounded at Sharpsburg; paroled at Appomattox Court House.

118 - Weathers, Thomas – enlisted as a private; absent without leave to the Farmville, Virginia hospital after June 19, 1863.

119 - Weems, H.M. – enlisted as a private; admitted to a Lynchburg, Virginia hospital with typhoid fever July 21, 1862; absent without leave after October 28, 1863.

120 - Weems, J.L. – enlisted as a private; present at Cedar Run, Second Manassas, and Sharpsburg; at hospital at Farmville, Virgina with the measles in October 1862; absent after leave after October 28, 1863.

121 - Yarbrough, G.M. – enlisted as a private; reduced in rank back to private circa July 1863; wounded at Chickamauga; absent without leave after November 28, 1863.

122 - Yarbrough, R.A. – enlisted as a private and promoted to First Sergeant; wounded at Cedar Run; in hospital recovering from the wounds until he returned to the regiment in October 1862; paroled at Appomattox Court House.

123 - Yarbrough, W.D. – enlisted as a private; died in Morristown, Tennessee February 4, 1864.

Company B - Raised in Blount and DeKalb Counties, Alabama and mustered at Warrenton Court House in Marshall County, Alabama on April 7, 1862.

1 - Baker, N.D. – enlisted as a private; captured during Lee's retreat

from Gettysburg at Falling Water July 14, 1863.

2 - Battles, A.J. – enlisted as a private; wounded at Sharpsburg and discharged due to heart disease.

3 - Battles, M.A. – enlisted as a private; killed at Cedar Run.

4 - Bearden, Samuel B. – enlisted as a private; died of typhoid fever in Richmond September 2, 1862.

5 - Brackett, Jesse J. – enlisted as a private; severely wounded in the leg at The Wilderness; paroled at Appomattox.

6 - Brinsfield, Zachariah – enlisted as a private; wounded at Second Manassas; Absent Without Leave after February 1864.

7 - Bristow, E.T. – enlisted as a private; died of pneumonia at American Hotel Hospital in Staunton, Virginia November 21, 1862.

8 - Broom, Issac – enlisted as a private; paroled at Appomattox Court House.

9 - Bryant, John W. – enlisted as a private; captured outside Knoxville, Tennessee January 22, 1864; sent as a Prisoner of War to Rock Island Barracks near Chicago, Illinois in February 1864 and released after the surrender June 20, 1865.

10 - Burgess, T.J. – elected Captain at formation; resigned from his post July 17, 1862 suffering from "hemorrhage in kidneys".

11 - Cain, R.W. – enlisted as a private and promoted to Assistant Surgeon/ Hospital Steward after the Suffolk Campaign May 10, 1863; paroled at Appomattox Court House.

12 - Carden, John L. – enlisted as a private.

13 - Chumley, George William – enlisted as a sergeant and rose to the rank of First Lieutenant June 27, 1864; furloughed home from July to October 1864.

14 - Clove, Samuel A. – enlisted as a private

15 - Clower, J.A. – enlisted as a private; killed at The Wilderness.

16 - Collier, David – enlisted as a private; absent without leave to the hospital in Richmond November 1862.

17 - Collier, Starling – enlisted as a private; detailed as a teamster; paroled at Appomattox Court House.

18 - Cooper, J.N. – enlisted as a private; died of typhoid fever February 20, 1863.

19 - Cothran, S.A. – enlisted as a private; deserted near Chattanooga, Tennessee August 27, 1863.

20 - Covington, John W. – enlisted as a private.

21 - Covington, Moses P. – enlisted as a private; wounded at Gettysburg, Pennsylvania.

22 - Cox, Samuel W. – elected Third Lieutenant at formation of company and eventually promoted to First Lieutenant; wounded at The Wilderness and dropped from the regiment due to poor health December 7, 1864.

23 - DeArmond, George W. – enlisted as a private; captured near Knoxville, Tennessee December 5, 1863; returned home after he took the oath on December 12, 1863.

24 - DeArmond, J.N. – elected Captain at formation of company; wounded at Gettys-burg; killed in action at The Wilderness.

25 - Ellis, F.M. – enlisted as a private; absent without leave on convalescents in a Richmond hospital November 8, 1864.

26 - Ellis, John W. – enlisted as a private; died of typhoid fever August 26, 1862.

27 - Furness, A.J. – enlisted as a private.

28 - Galaway, J.P. – enlisted as a corporal and reduced in rank to a private; deserted near Chattanooga, Tennessee.

29 - Galaway, W.H. – enlisted as private; deserted from the regiment near Chattanooga, Tennessee.

30 - Gaton, Silas – enlisted as a private and promoted to sergeant; wounded at Sharpsburg.

31 - Gilliland, David M. – enlisted as a private; sick in hospital most of the war.

32 - Gilliland, John R – enlisted as a private and promoted to sergeant.

33 - Gilliland, M.R. – enlisted as a private; paroled at Appomattox Court House.

34 - Gilliland, Nathaniel – enlisted as a sergeant; died of wounds November 30, 1863 near Knoxville, Tennessee.

35 - Gray, Thomas – enlisted as a private; died of disease at Camp Auburn April 24, 1862.

36 - Graham, M.V. – enlisted as a private; deserted from the regiment June 1, 1862.

37 - Graham, W.M. – enlisted as a private; died of disease at Camp Auburn May 23, 1862.

38 - Grilard, W. – enlisted as a private.

39 - Hagan, Joseph F. – enlisted as a private; captured at Big Shanty, Virginia June 11, 1864 and sent to Rock Island Barracks near

Chicago, Illinois.

40 - Hall, D.B. – enlisted as a private; suffered from rheumatism; spent most of 1864 sick at Chimborazo Hospital.

41 - Hall, John M. – enlisted as a private.

42 - Hammett, D.W. – enlisted as a private; died of pneumonia in Richmond, Virginia January 1, 1863.

43 - Hammett, Perry – enlisted as a private; captured near Knoxville, Tennessee December 3, 1863; sent to Rock Island Barracks near Chicago, Illinois as a prisoner of war.

44 - Hammett, Wilson – enlisted as a private; wounded at Fredericksburg, Virginia December 13, 1862.

45 - Hayes, Elisha – enlisted as a private; absent without leave sick at the hospital after May 8, 1863.

46 - Helton, C.A. – enlisted as a private; absent without leave sick after 1863.

47 - Higgins, Joseph – enlisted as a private; captured near Knoxville, Tennessee August 10, 1863; sent to the Rock Island Barracks near Chicago; released after taking the oath of allegiance December 13, 1863.

48 - Hill, William N. – enlisted as a private and promoted to sergeant; sick in hospital suffering from typhoid fever from July 21, 1862 to August 20, 1862; absent without leave in March 1863; returned to the regiment; paroled at Appomattox Court House.

49 - Homan, Isaac – enlisted as a private; sick at the hospital with rheumatism in 1862.

50 - Hooper, Isaac – enlisted as a corporal at the formation of the

company; deserted from the regiment November 2, 1863.

51 - Horton, J.H. – enlisted as a private; discharged from the service March 13, 1863.

52 - Horton, T.D. – enlisted as a private; killed in action near Knoxville, Tennessee November 26, 1863.

53 - Howell, Thomas – enlisted as a private; paroled at Appomattox Court House.

54 – Hughes, Joseph – enlisted as a private; wounded at Second Manassas and at Sharpsburg; discharged from regiment in October 1862.

55 - Jarrett, James J. – enlisted as a private; seriously wounded in action October 7, 1864; his medical records are a quite a lengthy volume themselves spanning from 1862-1865.

56 - Kay, Alexander A. – enlisted as a private; wounded at Cedar Run, Virginia and died of the wounds August 25, 1862.

57 - Kay, C.W. – enlisted as a private; sick in the hospital September 1862; paroled at a Appomattox Court House.

58 - Kay, John – enlisted as a private; absent sick without leave from May to August 1863.

59 - Kay, Silas – enlisted as a private; wounded at Chickamauga; paroled at Appomattox Court House.

60 - King, D.R. – elected First Lieutenant at formation of the company; wounded at Cedar Run; promoted to Captain September 15, 1862; died of wounds November 26, 1862 in Charlottesville, Virginia.

61 - King, R.D. – enlisted as a private.

62 - LeFoy, DeMarcus – enlisted as a private; captured near

Knoxville, Tennessee November 30, 1863; sent Rock Island Barracks near Chicago, Illinois.

63 - Lanford, Silas – enlisted as a private; paroled at Appomattox Court House.

64 - Lanford, Simeon – selected as company First Sergeant at formation of company; wounded near Knoxville, Tennessee November 25, 1863 and sent to Camp Chase, Ohio as a prisoner of war; transferred to Fort Delaware in N.Y.C.; released from there June 14, 1865.

65 - Lang, Ruben – enlisted as a private.

66 - Lankford, Elijah – enlisted as a private.

67 - Lankford, Andy – enlisted as a private; died of disease at Auburn, Alabama at formation of regiment.

68 - Lankford, Thomas – enlisted as a private; wounded and captured at Sharpsburg; exchanged; paroled at Appomattox Court House.

69 - Martin, J.B. – enlisted as a private; suffered from typhoid fever during July and August 1862; wounded severely May 23, 1864; absent without leave after August 1864.

70 - Mayfield, Hosea W. – enlisted as a private; sick in hospital from diarrhea most of 1863-64.

71 - Mays, James F. – enlisted as a private; killed in action at Gettysburg July 2, 1863.

72 - Mays, W.B. – enlisted as a private and promoted to corporal; paroled at Appomattox Court House.

73 - McBrayer, A.B. – enlisted as a private; killed in action at The

Wilderness.

74 - McBrayer, J.M. – enlisted as a sergeant; wounded at Gettysburg; died from his wounds at Martinsburg, Virginia July 9, 1863.

75 - McIntyre, J.C. – enlisted as a private.

76 - McMahon, J.C. – enlisted as a corporal; killed in action at Sharpsburg.

77 - Moon, Thomas L. – elected Second Lieutenant at formation of the company; resigned due to kidney disease; replaced by Jesse E. Ross.

78 - Morgan, Squire – enlisted as a private; sick in hospital most of 1862-63.

79 - Mullins, Columbus – enlisted as a private; died of disease July 27, 1862.

80 - Newman, Martin W. – elected Second Lieutenant; resigned October 6, 1862 suffering from chronic diarrhea.

81 - Owens, Bartley – enlisted as a private; died of "Phthisis Pulmonalis" (tuberculosis) February 6, 1863.

82 - Owens, B.F. – enlisted as a private; wounded and captured at Gettysburg; exchanged; absent without leave after March 1864; paroled at Talladega, Alabama May 29, 1865.

83 - Padgett, M.F. – enlisted as a private; discharged November 7, 1862 suffering from chronic dysentery.

84 - Payet, W.P. – enlisted as a private; died of disease June 17 1862.

85 - Payne, William – enlisted as a private; died before February 12, 1863.

86 - Pender, H.T. – enlisted as a private.

87 - Penn, H.T. – enlisted as a private; sick most of 1863; paroled at Appomattox Court House.

88 - Price, Bluford – enlisted as a private; admitted to Chimborazo Hospital in Richmond July 21, 1862 suffering from typhoid fever.

89 - Prince, Jackson H. – enlisted as a private; wounded at Second Manassas; deserted from the regiment November 2, 1862.

90 - Ray, Elisha – died of disease August 23, 1862.

91 - Roden, Thomas – enlisted as a corporal; paroled at Appomattox Court House.

92 - Rice, George W. – enlisted as a private; he was most likely wounded at Chickamauga; died in Harrietta, Georgia from disease January 26, 1864; he suffered from disease most of the war.

93 - Roden, Zealous – enlisted as a private.

94 - Ross, Jesse E. – elected as a Second Lieutenant at formation of the company; wounded in the foot at Cedar Run; resigned June 1, 1863.

95 - Sauls, F.M. – enlisted as a private; suffered from diarrhea during July and August 1862; absent without leave after February 14, 1864.

96 - Sauls, John – enlisted as a private; paroled at Appomattox Court House.

97 - Scott, Newton J. – enlisted as a private; died at Camp Auburn June 13, 1862.

98 - Segar, P.S. – enlisted as a private; killed in action at Cedar Run.

99 - Sheffield, Ervin – enlisted as a private; present most of the war; paroled at Appomattox Court House.

100 - Sheffield, P.F. – enlisted as a private; suffered from rheumatism and detailed to work at a hospital as an aid; deserted the hospital after getting sick April 13, 1863.

101 - Sheppard, John H. – elected Second Lieutenant at formation of the company; wounded at Gettysburg; promoted to First Lieutenant July 15, 1863; resigned January 22, 1864.

102 - Sheriff, William W. – enlisted as a private; died of disease August 30, 1862.

103 - Shirley, John Newton – enlisted as a sergeant and demoted to private; wounded at Gettysburg.

104 - Siggers, P.S. – enlisted as a private; killed in action at Cedar Run.

105 - Sitz, Isham – enlisted as a private; absent without leave after July 11, 1864.

106 - Sitz, James – enlisted as a private; absent without leave after April 26, 1863; returned to the regiment in time for the fight at Chickamauga; paroled at Appomattox Court House.

107 - Sitz, William R. – enlisted as a private; wounded in the left side at Gettysburg; sent home to furlough but stopped at a hospital.

108 - Stone, William H. – enlisted as a corporal and demoted to private; wounded in the shoulder at Gettysburg; sent to hospital to recover; absent without leave after August 30, 1864.

109 - Trotter, William – enlisted as a private.

110 - Vaughn, Robert – enlisted as a private; in hospital September

28, 1862.

111 - Windsor, O.C. – enlisted as a private; absent without leave after October 20, 1862.

112 - White, James – enlisted as a private; absent without leave after June 1, 1863; "deserted at leave lying in the mountains" during the march to Gettysburg.

113 - Wilbanks, W.S. – enlisted as a private; captured near City Point, Virginia April 12, 1865; ordered to be sent to Dunkirk, New York.

114 - Willsinger, J. – enlisted as a private.

115 - Wilson, A.W. – enlisted as a private; sick until death April 19, 1863.

116 - Wilson, J.W. – enlisted as a private; died of chronic diarrhea in Lynchburg, Virginia October 13, 1862.

117 - Windsor, O.C. – enlisted as a private; wounded at Chickamauga; absent without leave after October 1863.

118 - Works, James M. – enlisted as a private; wounded at Second Manassas; absent without leave after January 1, 1864.

119 - Yancey, Thomas B. – enlisted as a private and promoted to sergeant; sick at Farmville, Virginia hospital with typhoid fever July 21, 1862; wounded severely in the left leg October 7, 1864; leg amputated: "resection upper third of femur simple longitudinal incision over troch. major removing four inches"; he died from his wounds October 25, 1864.

120 - Yearta, W.L. – enlisted as a private; sick in Richmond hospital from October 28, 1862 to November 10, 1862 suffering from "meriosis of femur"; discharged November 26, 1862.

121 - Young, Thomas H. – enlisted as a private and promoted to corporal; sick in hospital from August 15, 1862 until November 18, 1862.

122 - Zimmerman, M.A. – enlisted as a private; captured April 6, 1865; sent to City Point, Virginia April 14, 1865.

Company C - Raised in Marshall County, Alabama and mustered at Warrenton Court House in Marshall County, Alabama April 7, 1862.

1 - Abston, D.L. – enlisted as a private; died of disease in Richmond, Virginia September 27, 1862.

2 - Bartlett, George W. – enlisted as a corporal at formation and rose to First Sergeant; paroled at Appomattox.

3 - Barton, David – enlisted as a private; captured at Gettysburg and remained a prisoner of war throughout 1864.

4 - Barton, W.J. – elected as Second Lieutenant at formation and received promotion to First Lieutenant; killed in action August 16, 1864.

5 - Bedford, James M. – elected First Lieutenant at formation; promoted to Captain.

6 - Bell, Wiley - enlisted as a private; wounded at Sharpsburg.

7 - Billingsley, J.W. – enlisted as a private; paroled at Appomattox Court House.

8 - Billingsley, H.A. – enlisted as a private; absent without leave after January 1863.

9 - Bishop, Enoch – enlisted as a private; absent without leave after May 1863.

10 - Bishop, William H. – enlisted as a private; died in Marion County, Tennessee August 22, 1863 from disease; his wife and brother Enoch were both present at his death.

11 - Bolin, James – enlisted as a private; deserted June 9, 1863.

12 - Bolin, Perry – elected sergeant at formation; deserted June 9, 1863.

13 - Boyle, J. – enlisted as a private; for an unknown reason, he was given 101 days of extra duty.

14 - Brasier, David – enlisted as a private.

15 - Burke, Henry – enlisted as a private.

16 - Cantrell, J.J. – enlisted as a private; absent without leave after June 1863.

17 - Carr, A.M.F. – enlisted as a private; spent most of the war in various hospitals suffering from chronic rheumatism.

18 - Carr, H.G. – enlisted as a private; wounded a Gettysburg.

19 - Carr, William – enlisted as a private; wounded during the siege of Suffolk, Virginia April 21, 1863.

20 - Childress, J. – enlisted as a private; discharged from service due to nephritis.

21 - Collins, Jerry A. – enlisted as a private; absent without leave after June 1863.

22 - Conn, S.A. – enlisted as a private and promoted to sergeant; absent without leave after March 1864 and found in the hospital suffering from "carbuncle".

23 - Cooper, J. – enlisted as a private.

24 - Cornelius, Madison – enlisted as a private; discharged because of diarrhea complications.

25 - Curbo, J.C. – enlisted as a private; absent without leave after The Wilderness.

26 - Denham, William – enlisted as a private; wounded in the right leg at Gettysburg; taken as a prisoner of war to David's Island in New York City.

27 - Dowlin, Berry – enlisted as a private; died of typhoid fever June 27, 1862.

28 - Dulin, William Benjamin – enlisted as a private; died of typhoid fever in Richmond June 27, 1862.

29 - Eberheart, Thomas – enlisted as a private; wounded severely in the soldier August 14, 1864; absent without leave after November 1864.

30 - Ellison, J.G. – enlisted as a private; wounded at Gettysburg.

31 - Ewing, Reuben T. – elected First Lieutenant and promoted to Captain; wounded at Second Manassas; paroled at Appomattox Court House.

32 - Fields, Berry – enlisted as a private; spent war in sick in the hospital.

33 - Franklin, James – enlisted as a private; died of typhoid fever June 27, 1862 in Wilmington, N.C. and is buried in a nearby cemetery.

34 - Furgeson, William – enlisted as a private; deserted from the regiment September 15, 1863.

35 - Gaines, N.P. – enlisted as a private; killed in action at Second Manassas.

36 - Galey, Hez. – enlisted as a private; wounded at Second Manassas.

37 - Garrard, James Moorin – enlisted as a private; captured at The

Wilderness.

38 - Garrard, John Franklin – enlisted as a private; wounded in the shoulder at The Wilderness.

39 - Gasaway, William – enlisted as a private; wounded at Sharpsburg; killed in action at Chickamauga.

40 - Gibbs, J.W. – enlisted as a private; absent without leave after July 1864.

41 - Gibson, Turner – enlisted as a private; died of typhoid fever November 16, 1862.

42 - Gilmore, Frances. – enlisted as a private.

43 - Goodwin, Samuel – enlisted as a private; died of pneumonia April 5, 1863.

44 - Gorce, H.P. – enlisted as a private; killed in action at Second Manassas.

45 - Gover, Hezekiah – enlisted as a private

46 - Graves, Aaron – enlisted as a private.

47 - Graves, J.T. – enlisted as a private at formation of company and promoted to sergeant; wounded at Chickamauga.

48 - Graves, William – enlisted as a corporal; wounded near Chattanooga October 15, 1863.

49 - Gregory, J.J. – enlisted as a private at formation of regiment and promoted to Second Lieutenant; wounded in the leg at The Wilderness and captured at Richmond April 3, 1865.

50 - Gregory, Thomas – enlisted as a private; wounded at Sharpsburg.

51 - Halcomb, Jesse – enlisted as a private; killed in action at Chickamauga.

52 - Harten, F. – enlisted as a private; wounded at Cedar Run.

53 - Higgins, J.L. – enlisted as a sergeant; demoted to private May 19, 1863; killed in action at Chickamauga.

54 - Hilton, J.H. – enlisted as a private; admitted to hospital August 9, 1863 suffering from "Val. Sclept; Shoulder's Flesh".

55 - Hobbs, Samuel – enlisted as a private; deserted September 15, 1863.

56 - Hunt, V.B. – enlisted as a private; deserted June 6, 1863.

57 - Hunter, E. – enlisted as a private.

58 - Hunter, M.V. – enlisted as a private; was from Walnut Grove (present-day Etowah County, Alabama); wounded on the middle finger of the right hand June 5, 1863; furloughed to recover.

59 - Johnson, F.M. – enlisted as a private; wounded and captured at Gettysburg; sent to Fort Delaware, N.Y.C. on October 15, 1863; died at Fort Delaware on November 8, 1863.

60 - Johnson, G.M. – enlisted as a private; discharged from service from compli-cations of diarrhea August 22, 1862.

61 - Johnson, John R. – enlisted as a private; wounded at Chickamauga.

62 - Johnson, Marion – enlisted as a private; wounded at Cedar Run and died of wounds August 14, 1862.

63 - Johnson, William – enlisted as a private; died in Richmond from disease October 28, 1862.

64 - Jones, John – enlisted as a private; died of disease in Richmond, Virginia July 3, 1863.

65 - Jones, W. – enlisted as a private; paroled at Talladega, Alabama May 24, 1865.

66 - Jordan, Benjamin Franklin – enlisted as a private.

67 - Kinnebrew, H.C. – elected Second Lieutenant at formation of company; promoted to Captain August 25, 1863; spent very little time sick and away from the company; he is one of the few original members of the regiment to be present at all major battles; paroled at Appomattox Court House.

68 - Lawing, Jethro – enlisted as a private; sick in the hospital during August and September 1862.

69 - League, William – enlisted as a private.

70 - Lee, W.A. – enlisted as a private; wounded at Second Manassas.

71 - Lee, Wiley – enlisted as a private; wounded at Second Manassas.

72 - Long, William S. – enlisted as a private; killed in action at Second Manassas.

73 - Mabry, P.J. – enlisted as a private; absent without leave after January 5, 1863.

74 - Martin, William – enlisted as a private; deserted September 15, 1863.

75 - McCluskey, D.W. – enlisted as a private and promoted to sergeant; killed in action at The Wilderness.

76 - McCracken, G.D. – enlisted as a private; sick at the hospital in October 1862; absent without leave from June to October 1863;

returned; paroled at Appomattox Court House.

77 - Miller, Andrew – enlisted as a private; worked as a mechanic and a shoemaker in Richmond, Virginia.

78 - Miller, Eldredge W. – enlisted as a private; wounded at Second Manassas and Sharpsburg.

79 - Miller, Jesse Freeman – enlisted as a private; present through July 1, 1863; absent without leave from July 1863 to the spring of 1864; paroled at Appomattox Court House.

80 - Miller, Needham E. – enlisted as a private; sick in hospital from November 1862 to July 1863; present thereafter; paroled at Appomattox Court House.

81 - Miller, T.T. – enlisted as a private.

82 - Morgan, A.W. – enlisted as a private; wounded at Chickamauga; died of wounds October 31, 1863 at a hospital in Harrietta, Georgia.

83 - Norris, J.A. – enlisted as a private; sick in hospital from July to November 1862; killed in action near Suffolk, Virginia April 20, 1863.

84 - Parrish, Isaac M. – enlisted as a private; killed in action at Cedar Run, Virginia.

85 - Parrish, James Madison – enlisted as a sergeant at formation of company; wounded at Second Manassas; wounded at Gettysburg July 2, 1863 (ball side and lung); sent to David's Island, N.Y.C. where he died of wounds July 26, 1863; buried at Cypress Hill National Cemetery.

86 - Parrish, Thomas Jefferson – enlisted as a sergeant at formation of company; paroled at Appomattox Court House; claimed to have

received at scalp wound at Second Manassas.

87 - Parrish, William A.T. – enlisted as a private; killed in action at Second Manassas.

88 - Pridmore, B. – enlisted as a private; died of convulsions in Lynchburg, Virginia November 8, 1862.

89 - Pruett, William – enlisted as a private; sick at hospital during October 1862; absent without leave after July 1864.

90 - Pugh, William Green – enlisted as a private; wounded at Cedar Run, Virginia; absent without leave after June 1863.

91 - Rape, James – enlisted as a private; killed in action at Cedar Run.

92 - Reed, Robert – enlisted as a private; wounded at Sharpsburg; died of disease November 19, 1862.

93 - Rice, B. – enlisted as a private; died in a Richmond hospital.

94 - Rodgers, Adolphus T. – enlisted as a private; died at Petersburg May 15, 1863.

95 - Rodgers, D. – enlisted as a private; suffered from rheumatism June 21, 1862.

96 - Rodgers, J.S. – enlisted as a private; killed in action at Chickamauga.

97 - Sampson, James – enlisted as a private; wounded at Cedar Run; wounded at Chickamauga; absent without leave after June 14, 1864.

98 - Sanders, Allen – enlisted as a private; present through most of the major battles; took the oath of allegiance February 14, 1865.

99 - Scott, Calvin – enlisted as a private; wounded seriously in the hand (lost two fingers) at Second Manassas; absent without leave

after June 15, 1863.

100 - Sligh, John – enlisted as a private; admitted to the hospital suffering from rheumatism October 1, 1862; it was discovered at this time that he was only seventeen years old; he was then furloughed home to recover later that month; returned to the regiment; killed in action at Spotsylvania Court House May 12, 1864.

101 - Sparks, Amon – enlisted as a private; died of measles at Auburn May 15, 1862.

102 - Sparks, Joshua A. – enlisted as a private; died of fever April 5, 1863.

103 - Sparks, J.T. – elected First Lieutenant at formation of company; admitted to hospital in October 1862 suffering from dysentery; wounded at Fredericksburg; reduced in rank to Private; resigned August 25, 1863.

104 - Sparks, Merida – enlisted as a private; absent sick at the hospital most of the major battles.

105 - Stallings, J.P. – enlisted as a private; absent sick at the hospital from May to October 1863; returned to the regiment and present until war's end; believed to have been paroled at Newton, North Carolina by the Tenth Michigan Cavalry about April 19, 1865.

106 - Staton, J.A. – enlisted as a private; killed in action at Second Manassas.

107 - Steglar, John – enlisted as a private; died of disease September 1, 1862 at Liberty Mills, Virginia.

108 - Thacker, H.A. – enlisted as a private; died of pneumonia in Charlottesville, Virginia May 1, 1864.

109 - Thornton, Joseph W. – enlisted as a private; died before

Joshua Price

August 22, 1863.

110 - Tippens, A. – enlisted as a private in company E and transferred to company C; sick in hospital with fever during July 1862; deserted from the regiment; took the oath of allegiance in Chattanooga, Tennessee August 11, 1864; was ordered by the Union army to remain north of the Ohio River.

111 - Trice, W.H. – enlisted as a private; died before February 14, 1863.

112 - Walker, William S. – elected Captain at formation of the company; resigned suffering from diarrhea October 1, 1862.

113 - Ward, Stokely Donaldson – enlisted as a private; deserted from the regiment September 19, 1863; took the oath of allegiance in Nashville, Tennessee April 10, 1865, in the presence of his family.

114 - Ward, T.B. – enlisted as a corporal; suffered from rheumatism during October 1862; deserted from the regiment September 19, 1863.

115 - Watts, Daniel Dodson – enlisted as a private in company E and later transferred to company C; wounded severely at Chickamauga: "gunshot wound right thigh and right scrotum".

116 - Watts, J.B. – enlisted as a private; paroled at Appomattox Court House.

117 - Winchester, William – enlisted as a corporal; wounded at Chickamauga.

118 - Winchester, Williamson – enlisted as a private; wounded at Sharpsburg; discharged from disease July 15, 1863.

Company D - Raised in Marshall County, Alabama and mustered at Warrenton Court House in Marshall County, Alabama April 7, 1862.

1 - Alexander, James – enlisted as a private and promoted to corporal.

2 - Anderson, J. – enlisted as a private.

3 - Anderson, Thomas J. – enlisted as a private; discharged due to disease in November 1862.

4 - Bailey, J.P. – elected Second Lieutenant at formation; resigned in 1862.

5 - Bailey, Peter – enlisted as a private; wounded at Chickamauga; killed at The Wilderness.

6 - Baldwin, E.B. – enlisted as a private and promoted to First Sergeant; wounded June 17, 1864 in his left leg; his leg was amputated at the thigh.

7 - Black, Calvin – enlisted as a private; absent during the summer of 1863; returned to the regiment for the Chickamauga Campaign and was present for the remainder of the war; paroled at Appomattox Court House.

8 - Black, Green – selected corporal at formation and promoted to sergeant; paroled at Appomattox Court House.

9 - Blankenship, R.J. – enlisted as a private; deserted June 9, 1863.

10 - Bowman, Vince P. – enlisted as a private; died of typhoid fever August 19, 1862.

11 - Buxton, H.H. – enlisted as a private; discharged October 9, 1862 by the surgeon because he was "under age".

12 - Corbin, James – enlisted as a private; died of typhoid fever at a Lynchburg, Virginia hospital October 30, 1862.

13 - Cardwell, James – enlisted as a private; killed in action at Gettysburg.

14 - Cardwell, Ruben – enlisted as a private; shot through the hip and through the thigh at Gettysburg where he was captured on July 5, 1863; sent to David's Island, New York as a prisoner of war. Killed while returning home from war.

15 - Chaney, G.W. – enlisted as a private; discharged due to suffering from chronic rheumatism.

16 - Cline, Mack W. – enlisted as a private; wounded and captured at Gettysburg; sent to David's Island in New York City where he recovered from his wounds; he was exchanged later that same year and returned to the regiment; he was severely wounded at The Wilderness and again August 16, 1864; the details of his wound were "fracture of the left fibula"; amputation of the limb was performed in Richmond the next day.

17 - Collins, W.W. – elected Second Lieutenant at formation of the company; wounded at Chickamauga and severely at The Wilderness; paroled at Appomattox Court House; present the entire war except while recovering from his wounds during May and June 1864.

18 - Coleman, G.W. – enlisted as a private; died of measles in Auburn May 29, 1862.

19 - Cox, Samuel A. – elected as Captain at formation of the company; died of typhoid fever April 28, 1863.

20 - Derbit, Michael – enlisted as a private.

21 - Doss, J.J. – enlisted as a private.

22 - Durden, D.L. – enlisted as a private; died at Petersburg, Virginia 02 AUG 1864.

23 - Eubanks, Thomas J. – enlisted as Adjutant and promoted to

Captain upon the death of Samuel A. Cox; was wounded at Lookout Valley, Tennessee October 28, 1863 and died the next day.

24 - Fielder, Charles P. – enlisted as a private; discharged due to kidney disease and fatigue August 1, 1862.

25 - Fielder, John – enlisted as a private and promoted to corporal.

26 - Franklin, Henry – enlisted as a private; died in Richmond of pneumonia March 6, 1863.

27 - French, J.D. – enlisted as a private.

28 - Gardner, R. – enlisted as a private.

29 - Gibson, William – enlisted as a private; wounded at Second Manassas.

30 - Gilbert, William – enlisted as a private and promoted to sergeant; wounded at Second Manassas; paroled at Appomattox Court House.

31 - Glidewell, J.V. – enlisted as a private.

32 - Gross, G.W. – enlisted as a private; paroled at Appomattox Court House.

33 - Gross, Jonathan – selected as First Sergeant at formation of company; reduced in rank June 30, 1863; wounded at Chickamauga.

34 - Haney, Oliver – enlisted as a private; suffered from rheumatism; deserted July 9, 1863.

35 - Harper, Allen – enlisted as a private; absent without leave after August 9, 1863.

36 - Haygood, Henry P. – enlisted as a sergeant; wounded and

captured at Sharpsburg; exchanged; wounded at Chickamauga.

37 - Hinds, J.M. – appointed Junior Second Lieutenant at formation of regiment; promoted Second Lieutenant; killed in action August 16, 1864.

38 - Howell, James Marion – enlisted as a private.

39 - Hubbard, F.M. – enlisted as a private; died in the hospital from typhoid fever August 6, 1862.

40 - Hughes, W.N. – enlisted as a corporal and demoted to private; absent without leave after April 1864.

41 - Jones, William C. – enlisted as a corporal; died of disease at Auburn June 8, 1862.

42 - Jordan, J.J. – enlisted as a private; died of diarrhea November 1, 1862.

43 - King, James W. – enlisted as a private; sick at the hospital suffering from rheumatism in October 1862; sick most of 1863 and early 1864; seriously wounded July 15, 1864.

44 - King, W.D. – enlisted as a sergeant and demoted to private; wounded in the shoulder at The Wilderness and left on the battlefield.

45 - Long, J.A. – enlisted as a private.

46 - Lloyd, J. – enlisted as a private; paroled at Talladega, Alabama May 20, 1865.

47 - Lyles, J.M. – enlisted as a private; wounded at Gettysburg.

48 - Martin, D.C. – enlisted as a private; present through Gettysburg; absent after Chickamauga.

49 - Martin, J.B. – enlisted as a private and promoted to corporal.

50 - McDermont, J.L. – enlisted as a private; wounded at Chickamauga September 19, 1863 and died the next day.

51 - Millirous, M.L. – enlisted as a private; wounded at Gettysburg July 2, 1863; sent to David's Island, N.Y.C.; exchanged in late 1863; returned to regiment; sick during most of 1864.

52 - Mitchell, Seaborne J. – enlisted as a private; seriously wounded in the face at Second Manassas; lost an eye; spent the rest of 1862 and also 1863 recovering; absent without leave after 1864.

53 - Place, Emanuel – enlisted as a private; discharged because of disease in September 1862.

54 - Prescott, Anderson – enlisted as a private; paroled at Augusta, Georgia May 22, 1865.

55 - Putman, William – enlisted as a private and promoted to sergeant; court martialed by the regiment October 28, 1862, but the papers are illegible; wounded at Chickamauga; paroled at Appomattox Court House.

56 - Raulan, P.B. – enlisted as a sergeant.

57 - Renfro, B.H. – enlisted as a private; paroled at Appomattox Court House.

58 - Rice, D.W. – enlisted as a private.

59 - Ridgeway, J.S. – elected Second Lieutenant at formation of company; promoted to First Lieutenant February 20, 1863; wounded and captured at Gettysburg; sent from Harrisonburg, Pennsylvania military prison to Fort Delaware October 31, 1863; was exchanged after this transfer.

60 - Rodgers, Elisha – enlisted as a private and promoted to

corporal; captured August 14, 1864; exchanged from Point Lookout, Maryland March 15, 1865.

61 - Romine, Thomas – enlisted as a private; died of disease April 7, 1863.

62 - Rowan, T. – enlisted as a private.

63 - Sanders, John – enlisted as a private; killed in action at Second Manassas.

64 - Saunders, John – enlisted as a private; killed in action at Cedar Run.

65 - Sauter, Daniel – enlisted as a private; deserted from the regiment on the Chickamauga September 19, 1863.

66 - Sauter, Edward – enlisted as a private; killed in action at Cedar Run.

67 - Scruggs, James – enlisted as a private; present through most of the major battles; paroled at Appomattox Court House.

68 - Scruggs, Robert Milton – enlisted as a private and promoted to corporal; wounded at Cedar Run; present through most of the major battles; paroled at Appomattox Court House.

69 - Selvage, Tom – enlisted as a sergeant; died sometime before February 12, 1863.

70 - Smallwood, F.M. – enlisted as a private; worked as a teamster.

71 - Smith, J.C. – enlisted as a private; absent without leave after February 17, 1864.

72 - Smith, James M. – elected Second Lieutenant at formation of the company; resigned due to a broken ankle June 13, 1862.

73 - Smith, Joel D. – enlisted as a private; wounded at Chickamauga.

74 - Smith, P.A. – enlisted as a private; wounded and captured at Gettysburg; sent to Fort Delaware as a prisoner of war; released June 14, 1865,

75 - Smith, W.W. – enlisted as a private; killed in action at Chickamauga.

76 - Stephens, S.P. – enlisted as a private; captured near Knoxville, Tennessee December 1, 1863; taken as a prisoner of war to Louisville, Kentucky.

77 - Stewart, John – enlisted as a private; wounded at Cedar Run; sent home to recover from wounds; died at his home in Marshall County, Alabama from wounds March 25, 1863.

78 - Stewart, William Milton – enlisted as a private; missed the 1862 campaign sick in Danville, Virginia suffering from typhoid fever; returned for the 1863 and 1864 campaigns.

79 - Tidwell, A.J. – enlisted as a corporal.

80 - Tipton, S.C. – enlisted as a private; discharged from chronic rheumatism in his right arm December 24, 1862.

81 - Turner, A.J. – enlisted as a sergeant and demoted to private; absent without leave after April 15, 1864.

82 - Walker, John – enlisted as a private; wounded at Sharpsburg; reduced in ranks from corporal "by sentence of regimental court martial, August 8, 1864".

83 - Winder, J.K. – enlisted as a private and promoted to corporal; sick at hospital during the fall of 1862; paroled at Appomattox Court House.

84 - Williams, M.L. – enlisted as a private; wounded in the thigh at

Gettysburg; sent to Fort Wood in New York City; died of disease April 15, 1864.

85 - Wilson, John – enlisted as a private; died of disease in Lynchburg, Virginia September 4, 1862.

86 - Wilson, L.W. – enlisted as a private; wounded at Chickamauga; absent without leave after November 1863.

87 - Wood, H.W. – enlisted as a private; in hospital with diarrhea during September 1862; absent without leave after October 25, 1862.

88 - Young, W.P. – enlisted as a private; sick in hospital after August 2, 1863.

Company E - Raised in Marshall County, Alabama and mustered at Warrenton Court House in Marshall County, Alabama on April 7, 1862.

1 - Amos, James W. – enlisted as a private and was promoted to sergeant; surrendered, but not at Appomattox Court House.

2 - Andrews, Oliver P. – enlisted as a private.

3 - Appleton, Thomas W. – enlisted as a private at formation and earned commission as First Lieutenant; was originally a member of Company G, but transferred to Company E; paroled at Appomattox Court House.

4 - Armstrong, James – enlisted as a private at formation of company; deserted from the regiment in early 1863.

5 - Arnold, James F. – enlisted as a private; was originally a member of Company G but was transferred to Company E; died of disease in June 1863.

6 - Atwood, Lemuel G. – enlisted as a private and promoted to sergeant.

7 - Bailey, J.S. – enlisted as a sergeant at formation.

8 - Barnes, George W. – enlisted as a private at formation.

9 - Benson, James S. – enlisted as a sergeant; deserted October 4, 1863.

10 - Benson, Spencer – enlisted as a sergeant; killed in action at Cedar Run.

11 - Benson, Wiley A. – enlisted as a private; deserted October 4, 1863.

12 - Busby, James W. – enlisted as a private; wounded at Cedar Run, Virginia.

13 - Cash, Peter – enlisted as a private; transferred from Company G to Company E; wounded in the leg at Gettysburg; sent to David's Island in New York City as a prisoner of war where he died of his wounds September 1, 1863; he is buried at Camp Hill Cemetery in grave #833.

14 - Chambers, Spencer – enlisted as a private; deserted in 1862.

15 - Clark, John – enlisted as a private; wounded in the "pinky" finger at The Wilderness and had to have it amputated.

16 - Clark, John R. – enlisted as a sergeant at formation.

17 - Cook, Stephen – enlisted as a private; was company musician; wounded at Gettysburg.

18 - Corbin, William Riley – enlisted as a private; absent without leave after April 1863.

19 - Crossack, Peter – enlisted as a private; wounded in the shoulder and captured at Gettysburg; exchanged in September 1863; absent without leave after January 1, 1864.

20 - Cunningham, A.V. – enlisted as a private; wounded at Sharpsburg.

21 - Cunningham, Calvin – enlisted as a private; wounded at Cedar Run; paroled at Lynchburg, Virginia April 14, 1865.

22 - Cunningham, George – enlisted as a private and promoted to sergeant; deserted from the regiment October 1, 1863 and captured by Union forces at Larkinsville, Alabama October 30, 1863; was taken to Camp Morton in Indianapolis, Indiana where he pled to take the oath so he could go home; he died at Camp Morton of "inflammation of the lungs" February 11, 1865 and was buried in Greenlawn Cemetery in grave #130.

23 - Cunningham, H. – enlisted as a private; died of disease in Staunton, Virginia October 4, 1862.

24 - Davis, George W. – enlisted as a corporal.

25 - Dickson, William J. – enlisted as a private.

26 - Dixon, J. – enlisted as a private; died of typhoid fever in Lynchburg, Virginia August 4, 1862.

27 - Dobbs, Jerry M. – enlisted as a private; sick in hospital October 1862.

28 - Dove, James P. – enlisted as a corporal; suffered from smallpox in fall 1862.

29 - Dunn, Samuel C. – enlisted as a private; absent without leave after January 13, 1863.

30 - Eason, Bryant C. – enlisted as a private; deserted from the regiment in May 1863.

31 - Farriss, Joseph – enlisted as a private; deserted August 2, 1863.

32 - Fields, H.B. – enlisted as a private.

33 - Fletcher, L.T. – enlisted as a private; died of dysenteria December 25, 1863.

34 - Garmon, Albert – enlisted as a private; promted to First Lieutenant.

35 - Gibbs, William J. – enlisted as a private; killed in action at Sharpsburg.

36 - Gibson, Abner C. – enlisted as a private; deserted September 19, 1863.

37 - Gibson, George B. – enlisted as a private.

38 - Gilbert, J.J. – enlisted a private; promoted to corporal.

39 - Gilbert, W.F. – enlisted as a private; wounded at Second Manassas.

40 - Gilbert, W.W. – enlisted as a private; wounded during the siege of Suffolk May 1, 1863; paroled at Appomattox Court House.

41 - Glazier, Thomas V. – enlisted as a private.

42 - Godsey, George W. – enlisted as a private; discharged from service in 1862.

43 - Hall, Jacob E. – enlisted as a private; sick in Richmond hospital after May 1863.

44 - Hammonds, Hampton H. – enlisted as a private; discharged of tuberculosis from Lynchburg, Virginia hospital November 29, 1862. Was a mechanic in his civilian occupation.

45 - Harris, George – enlisted as a private; wounded and captured at

Joshua Price

Sharpsburg; died as a prisoner of war September 27, 1862.

46 - Harris, Thomas B. – enlisted as a private.

47 - Hays, Oliver P. – enlisted as a private; killed at Second Manassas.

48 - Hickman, Joseph – enlisted as a private; wounded at Second Manassas; absent without leave after March 1864.

49 - Horton, S.E. – enlisted as a private; wounded severely at Chickamauga; sent to a hospital in Newman, Georgia where he died of his wounds November 16, 1863.

50 - Johnson, C. – enlisted as a private.

51 - Johnson, W.C. – enlisted as a private; died of disease in Lynchburg, Virginia October 28, 1862.

52 - Johnson, W.P.H. – enlisted as a private; deserted in 1862.

53 - Jones, Daniel – enlisted as a private and promoted to corporal; deserted in October 1863.

54 - Jones, William M. – enlisted as a private; suffered from typhoid fever in September 1862; missing at muster roll of March 1863 and declared dead in August 1863; found to have been discharged from a hospital in September 1864 (date of his discharge from the hospital is not known).

55 - Jordan, Harriet E. – enlisted as a private.

56 - Kennedy, James K. – enlisted as a private.

57 - Kennemer, Levi – elected Second Lieutenant at formation of the company; perhaps lost his commission with the consolidation of companies G and E.

58 - Knight, Henry – enlisted as a private; wounded at Cedar Run;

absent without leave thereafter.

59 - Knight, J.T. – enlisted as a private; wounded at Sharpsburg 17 SEPT 1862; absent without leave September 20, 1862; admitted to a Richmond hospital October 5, 1862; transferred to a hospital in Alabama.

60 - Knight, S.G. – enlisted as a private; absent without leave after September 1862; died of pneumonia at a hospital in Danville, Virginia March 3, 1863.

61 - Kuykendall, Absolum – enlisted as a private; killed in action at Sharpsburg.

62 - Lackey, C.R. – enlisted as a corporal; deserted from the regiment in May 1863.

63 - Lacy, Allen – enlisted as a private.

64 - Lacy, J.T. – enlisted as a private.

65 - Lasely, William H. – enlisted as a private; promoted to corporal June 15, 1863; wounded in the hand and lung at Gettysburg July 2, 1863; sent to Fort Delaware, N.Y.C. where he died from his wounds September 1, 1863.

66 - Law, Robert C. – enlisted as a private; wounded at Chickamauga.

67 - Lee, John K. – enlisted as a private.

68 - Lewis, Tarleton – enlisted as a private; died of disease at a hospital September 11, 1862.

69 - Mathews, D.W. – enlisted as a private; wounded at Gettysburg.

70 - McAdams, H.D. – enlisted as a private; wounded at The

Joshua Price

Wilderness; died of wounds May 19, 1864.

71 - Morgan, W.F. – enlisted as a private; died of typhoid fever September 12, 1862.

72 - Morris, W.D. – enlisted as a sergeant at formation of company; paroled at Appomattox Court House.

73 - Morris, W.S. – enlisted as a private; wounded at Second Manassas; furloughed home to recover from illness in the early months of 1863; returned to the regiment in May 1863.

74 - Mosley, D.B.F. – enlisted as a private; died in Danridge, Tennessee January 16, 1864.

75 - Nunnelley, Bedford – enlisted as a private; deserted August 16, 1862; paroled May 25, 1865.

76 - Peace, Jesse – enlisted as a private; wounded at Cedar Run; died from wounds at Gordonsville, Virginia August 14, 1862.

77 - Pendergrass, Landon – enlisted as a private; worked as a teamster; killed in action at Gettysburg July 2, 1863.

78 - Pendergrass, P.H. – enlisted as a private; died of typhoid fever in Lynchburg, Virginia August 11, 1862.

79 - Pendergrass, Willie – enlisted as a musician.

80 - Phillips, C.W. – enlisted as a private; killed at Second Manassas.

81 - Phillips, Henry J. – selected as First Sergeant at formation of company; demoted to private late 1863; absent without leave after January 1, 1863.

82 - Proctor, M. – enlisted as a private; sick at hospital after October 1862.

83 - Ragsdale, George W., Jr. – enlisted as a corporal.

84 - Ragsdale, George W., Sr.. – enlisted as a private; suffered from pneumonia March 9, 1863; killed at Gettysburg.

85 - Ragsdale, S.C. – enlisted as a private; killed at Chickamauga.

86 - Rayburn, Samuel K. – elected Captain at formation of the company; resigned June 13, 1862.

87 - Rector, Mark – enlisted as a private in company G; wounded at Chickamauga; paroled at Appomattox Court House.

88 - Reed, George – enlisted as a private; died of fever February 1, 1863.

89 - Reno, W.W. – enlisted as a private in company G; later transferred to company E; wounded at Chickamauga; sent to guard ironworks around Richmond in July 1864.

90 - Roberts, L.F. – enlisted as a private; absent without leave after June 1, 1863; deserted and captured by Union forces in December 1863; given the oath of allegiance at Nashville, Tennessee April 8, 1865.

91 - Robinson, W.B. – enlisted as a private; wounded in the head at Sharpsburg; discharged March 23, 1863.

92 - Roden, Emory – enlisted as a private; deserted from the regiment and surrendered himself to the First Alabama Cavalry at Paint Rock Bridge, Alabama; he was "desirous of taking the oath of allegiance".

93 - Roden, Francis K. – enlisted as a private in company G; later transferred at consolidation of E and G companies; deserted 1862.

94 - Roden, John B. – enlisted as a private; deserted 1862.

95 - Roe, John K. – enlisted as a private; absent without leave after May 10, 1863.

96 - Ross, John – elected Second Lieutenant; kept his rank through the consolidation of companies E and G.

97 - Ross, F.M. – elected First Lieutenant at formation of company; resigned June 1, 1862.

98 - Rucks, G.W. – enlisted as a private in company G; transferred at consolidation with company E; wounded at Cedar Run; furloughed home January 1, 1863.

99 - Sampson, J. – enlisted as a private.

100 - Sanders, J.A. – enlisted as a private.

101 - Scott, C. – enlisted as a private; wounded at Cedar Run.

102 - Shaw, Charles – enlisted as a private and promoted to sergeant; wounded in the arm at Gettysburg; sent to Point Lookout, Maryland as a prisoner of war; his wounded arm was amputated and he died there July 20, 1863.

103 - Simpson, S.V. – enlisted as a corporal in company G; died of typhoid fever in Lynchburg, Virginia September 3, 1862.

104 - Sly, J.T. – enlisted as a private.

105 - Small, J.B. – elected Second Lieutenant at formation of company G; elected First Lieutenant at consolidation of companies G and E; elected Captain of company E June 1, 1863; died in White Plains, Alabama June 24, 1864.

106 - Smelser, Adam – enlisted as a private and promoted to sergeant; deserted from the regiment October 21, 1863.

107 - Smith, A.H. – enlisted as a private; wounded at Second

Manassas; died of wounds November 27, 1862.

108 - Smith, J.W. – enlisted as a corporal and demoted to private.

109 - Sotherland, Adam – enlisted as a private; wounded at Chickamauga and given furlough home.

110 - Sotherland, A.S. – enlisted as a sergeant and demoted to private; wounded at Second Manassas; absent without leave after August 10, 1863.

111 - Sotherland, Harvey – enlisted as a private; wounded at Cedar Run; furloughed home to recover.

112 - Sutton, James H. – enlisted as a corporal.

113 - Taylor, John Dykes – promoted to regimental Ordnance Sergeant; absent without leave after The Wilderness. First regimental historian.

114 - Tennyson, John P. – enlisted as a private and promoted to corporal; wounded at Sharpsburg; wounded accidentally in Hagerstown July 2, 1863; wounded in the right shoulder at The Wilderness and again in the shoulder August 16, 1864.

115 - Thornburg, A.B. – enlisted as a private; died of disease August 11, 1862.

116 - Thornburg, A.W. – enlisted as a private; died in Lynchburg, Virginia from disease October 1, 1862.

117 - Thornburg, Charles W. – enlisted as a private; died before February 12, 1863.

118 - Turner, W.C. – enlisted as a private; deserted in 1862.

119 - Tyler, Spencer C. – enlisted as a private; promoted to First

Lieutenant July 17, 1862; resigned June 1, 1863; died in White Plains, Alabama June 24, 1864.

120 - Waller, Thomas P. – enlisted as a private; wounded in the left femur at Gettysburg; recovered at a Charlottesville hospital; paroled at Appomattox Court House.

121 - Ward, Alfred D. – enlisted as a private; became sick in the fall of 1862 and sent to Charlottesville; he was forced to stay in Charlottesville because all cars to Richmond were full; he died in Charlottesville of disease December 18, 1862.

122 - Ward, J.T. – enlisted as a private.

123 - Warley, J.W. - enlisted as a private in company C and transferred to company E; absent without leave after June 1863.

124 - Watkins, William – enlisted as a private; wounded in the shoulder at Sharpsburg and captured; he was apparently released or exchanged; recovered in a Richmond hospital.

125 - Watts, J.J. – enlisted as a private; died of disease in Richmond, Virginia July 7, 1862.

126 - Wilemon, Benjamin E.K. – appointed musician at formation of company.

127 - Wilemon, John B. – enlisted as a private.

128 - Wilkes, Richard – appointed sergeant and demoted to private; absent without leave after September 17, 1863; surrendered to the First Alabama Cavalry at Paint Rock Bridge; wanted to take the oath of allegiance.

129 - Williams, William – enlisted as a private; deserted from the regiment; captured by Union soldiers in November 1863; taken by Union soldiers to Washington, D.C. as a prisoner of war; forwarded to Philadelphia, Pennsylvania.

130 - Willimore, T. – enlisted as a private; sick in Charlottesville hospital suffering from typhoid fever November 21, 1862.

131 - Winkles, Jackson – enlisted as a private; large sick record.

132 - Winkles, Richard – enlisted as a private and promoted to corporal; attached to the "pioneer corps"; wounded in the abdomen at The Wilderness; paroled at Appomattox Court House.

133 - Woods, John – enlisted as a private; deserted October 25, 1863.

134 - Yearta, Berry – enlisted as a private in company G; killed in action at Chickamauga.

Company F - Raised in Blount County, Alabama and mustered at Warrenton Court House in Marshall County, Alabama April 10, 1862.

1 - Adrian, James F. – enlisted as a private at formation of company and earned a commission as First Lieutenant; wounded at Gettysburg July 2, 1863.

2 - Anderson, A. – enlisted as a private.

3 - Anderson, F.M. – enlisted as a private; prisoner of war.

4 - Ballard, Louis – enlisted as a private.

5 - Ballard, Daniel J. – enlisted as a private; captured while fighting around Knoxville, Tennessee.

6 - Ballard, Sherrod – enlisted as a private; killed in action at Second Manassas.

7 - Barnes, R.C. – enlisted as a private; absent without leave after

Joshua Price

June 1862; he was never engaged in combat.

8 - Beavers, Elbert – enlisted as a private and promoted to corporal.

9 - Benefield, J.M.D. – enlisted as a private; wounded at Chickamauga and later captured at Deep Bottom.

10 - Bennett, J.F. – enlisted as a sergeant; was wounded and captured at Gettysburg; exchanged; was bussed down in rank to private after extended absence without leave June 1, 1864.

11 - Bentley, Richard – enlisted as a private; spent from July to December 1862 in the hospital suffering from typhoid fever; declared a deserter May 4, 1863 following the Suffolk Campaign; was present for duty during the summer campaign of 1864; paroled at Appomattox Court House.

12 - Bentley, R.P. – enlisted as Fourth Corporal; absent without leave after May 1864.

13 - Bland, Milton – enlisted as a private; suffered from both hepatitis and bronchitis and spent his entire service recuperating in the hospital; died of "sundries" in the hospital April 8, 1863.

14 - Bland, William – enlisted as a private; died of measles May 31, 1862 in Auburn.

15 - Bland, William J. – enlisted as a private; was the father of Milton and William Jr.; deserted July 18, 1862 and never saw combat.

16 - Blann, M. – enlisted as a private.

17 - Boyd, Joseph J. – enlisted as a private; discharged from the service in October 1862 from disease (recommendation of Major John Wiggonton).

18 - Bradley, Joel – enlisted as a private; died of measles at a Lynchburg, Virginia hospital October 27, 1862.

19 - Brock, B.M. – enlisted as a sergeant; discharged December 22, 1862 due to "organic disease of the heart".

20 - Burgess, William – enlisted as a private.

21 - Burns, James M. – enlisted as a private; captured while fighting around Knoxville, Tennessee December 5, 1863; sent to Louisville, Kentucky and then to Rock Island, Illinois as a prisoner of war.

22 - Burns, G.W. – enlisted as a private; discharged from the army October 5, 1862 from complications of diarrhea.

23 - Byrd, Henry – enlisted as a private.

24 - Candle, Berry T. – enlisted as a private; discharged from the army September 15, 1862 suffering from organic liver disease.

25 - Candle, William – enlisted as a private; captured during the Suffolk Siege April 16, 1863; exchanged; wounded at Chickamauga.

26 - Coffey, S.D. – selected as a sergeant at formation of the company; sick most of 1863.

27 - Cone, Derrill – enlisted as a private; wounded at Second Manassas; discharged from service in April 1863.

28 - Cook, Robert – enlisted as a private; captured at Gettysburg July 4, 1863 and sent to Fort Delaware in New York City as a prisoner of war; died of diarrhea at Hammond General Hospital in Point Lookout, Maryland December 3, 1863.

29 - Cook, Stephen – enlisted as a private; was one of the company musicians; wounded at Gettysburg.

30 - Coppack, John J. – enlisted as a private; captured at Port Royal, Virginia August 28, 1863; exchanged; died June 7, 1864 at his home

in Blountsville, Alabama.

31 - Cornelius, William J. – enlisted as a private; paroled at Appomattox Court House.

32 - Crawford, W.T. – enlisted as a private and promoted to sergeant; suffered from diarrhea and in the hospital June 21, 1862; paroled at Appomattox Court House.

33 - Crindle, William J. – enlisted as a private; captured April 16, 1865 and paroled in Macon, Georgia May 2, 1865.

34 - Crossack, J.J. – enlisted as a private; captured at Port Royal September 6, 1863; exchanged April 37, 1864.

35 - Hamilton, William F. – enlisted as a private; wounded and captured at Gettysburg July 2, 1863; sent to Fort Delaware as a prisoner of war; died of rheumatism October 16, 1863.

36 - Harris, Andrew Jackson – enlisted as a private; discharged suffering from rheumatism July 25, 1862.

37 - Harper, H.S. – enlisted as a private.

38 - Harrison, B. – enlisted as a private; discharged suffering from rheumatism October 11, 1862.

39 - Hash, A.S. – enlisted as a private.

40 - Hayes, B.F. – enlisted as a private; absent without leave after June 20, 1863.

41 - Henderson, Clinton – enlisted as a private and promoted to sergeant; wounded at Gettysburg July 2, 1863 and never returned to the regiment.

42 - Henderson, J.D. – enlisted as a private; wounded severely in the leg at The Wilderness.

43 - Henderson, T.P. – enlisted as a private; wounded at Chickamauga.

44 - Hinds, R.W. – elected Second Lieutenant at formation of the company; resigned his commission because of rheumatism July 14, 1862.

45 - Huffstutler, G.W., Sr. – elected Second Lieutenant at formation of company; resigned his commission due to age and disease in May 1862; died of measles June 1, 1862.

46 - Huffstutler, Lewis – enlisted as a private and promoted to corporal; killed at Chickamauga.

47 - Ingle, Daniel – enlisted as a corporal; sick 1863-64.

48 - Ingram, D. – enlisted as a private.

49 - Johnson, Jarrett J. – enlisted as a private; sick at the hospital October 16, 1862; killed in action at Chickamauga .

50 - Jordan, J.D. – enlisted as a private; paroled at Talladega, Alabama May 29, 1865.

51 - Keel, H.J. – enlisted as a private; sent home sick from Chimborazo Hospital in Lynchburg, Virginia October 25, 1862.

52 - Kirk, S. – enlisted as a private.

53 - Lambert, D.C. – enlisted as a private; captured at Spotsylvania Court House on May 9, 1864; exchanged from Point Lookout, Maryland February 13, 1865.

54 - Lambert, John C. – enlisted as a private; wounded at Second Manassas and again at Sharpsburg; sick at hospital during October 1862; paroled at Appomattox Court House.

55 - Little, J.P. – enlisted as a private; died from complications of diarrhea August 25, 1862.

56 - Matthews, Levi P. – enlisted as a private; died from "inflammation of lungs" January 26, 1864.

57 - McAnnally, J.C. – enlisted as a private; absent without leave after June 1862.

58 - McCoy, John – enlisted as a private; wounded at Chickamauga.

59 - McCulchen, J.D. – enlisted as a private; wounded and captured at Gettysburg.

60 - McHan, James – enlisted as a private; wounded and captured at Gettysburg July 2, 1863 and his leg was amputated at the thigh.

61 - McMahon, John – enlisted as a private; killed at Sharpsburg.

62 - McMurray, Alexander – enlisted as a private; wounded at Second Manassas; wounded and captured at Gettysburg; sent as a prisoner of war to Fort Delaware; died in captivity of chronic diarrhea February 1, 1865.

63 - McMurray, M.C. – enlisted as a private; wounded and captured at Gettysburg; sent to Fort Delaware as a prisoner of war.

64 - McPherson, B. – enlisted as a private; died of typhoid fever at Camp Auburn June 10, 1862.

65 - McPherson, Elijah – appointed sergeant at formation of the company; died from typhoid fever in Blount County, Alabama July 5, 1862.

66 - Miller, Jasper L. – enlisted as a private; discharged from disease July 28, 1862.

67 - Miller, L.O. – enlisted as a musician; paroled at Appomattox Court House.

68 - Owens, Robert – enlisted as a private.

69 - Parks, Andrew H. – enlisted as a private; died of typhoid fever August 1, 1862.

70 - Patterson, J.P. – enlisted as a private and promoted to sergeant; paroled at Appomattox Court House.

71 - Ratliff, John – enlisted as a private; absent without leave after April 18, 1864.

72 - Ratliff, Joshua – enlisted as a corporal.

73 - Ratliff, Levi – enlisted as a private; died of chronic diarrhea July 3, 1863.

74 - Reno, Benjamin A. – enlisted as a private; absent without leave after July 20, 1862; captured by Union soldiers as a "rebel deserter"; sent to Louisville Military Prison in August 1864.

75 - Rhodes, William H. – enlisted as a private; sick in hospital suffering from rheumatism October 10, 1862; discharged from service September 29, 1862.

76 - Roberts, J.C. – enlisted as a corporal.

77 - Roberts, J.P. – enlisted as a private; reduced in rank from a Second Lieutenant by the examining board August 25, 1863; deserted September 18, 1863.

78 - Scott, L.A. – enlisted as a private; paroled at Montgomery May 15, 1865.

79 - Self, Henry – enlisted as First Sergeant; died of typhoid fever in Richmond July 3, 1862.

80 - Self, Jesse – enlisted as a private; detached to provost guard at Lynchburg, Virginia in June 1864.

81 - Self, John L. – enlisted as a private and promoted to sergeant; wounded at Cedar Run; was wounded in the right leg on August 16, 1864; had a long history of illness during the war.

82 - Shannon, M.L.M. – enlisted as a sergeant and demoted to private; wounded at Cedar Run; absent without leave after November 1, 1863.

83 - Sims, A.J. – enlisted as a private; absent sick after May 1, 1863.

84 - Sims, James D. – enlisted as a private; took oath of allegiance at Point Lookout, Maryland June 19, 1865; long history of illness during the war.

85 - Sims, John D. – enlisted as a private; deserted July 18, 1862.

86 - Sims, John J. – enlisted as a private and promoted to corporal; wounded at Cedar Run.

87 - Sinclair, Thomas J. – elected Second Lieutenant at formation of company; paroled at Appomattox Court House.

88 - Sinley, S.A. – enlisted as a private.

89 - Smith, James – enlisted as a corporal; wounded and captured at Gettysburg; taken to Fort Delaware where he died December 5, 1863 from "inflammation of lungs"; buried at "Jersey Shore opposite port".

90 - Smith, J.L. – enlisted as a private; spent much time in hospital during the war; wounded in the hand October 7, 1864.

91 - Smith, Matterson – enlisted as a private; wounded in the eye at Sharpsburg; was declared a deserter but discharged October 6, 1862 after being found in a hospital.

92 - Smith, T.A. – enlisted as a private; wounded at Second Manassas; paroled at Appomattox Court House.

93 - Staton, G.D. – enlisted as a private; captured at Gettysburg; sent to Fort McHenry, Maryland as a prisoner of war and then to Fort Delaware; exchanged September 17, 1863 and sent to City Point, Virginia; wounded seriously in the right leg August 16, 1864; the wounded leg was amputated the next day; by December 31, 1864 he was fully recovered and retired January 10, 1865.

94 - St. John, Colombus B. – elected Second Lieutenant at formation of the company; promoted to Major; wounded in the thigh at Gettysburg; retired from service on November 2, 1864.

95 - Stroup, M.J. – enlisted as a private; admitted to Danville, Virginia hospital September 25, 1862 suffering from typhoid fever; died in Bristol, Tennessee on April 13, 1864.

96 - Thompson, George – enlisted as a private; died of measles at Camp Auburn on May 24, 1862.

97 - Vincent, John – enlisted as a private; wounded and captured at Sharpsburg; exchanged; sick in hospital October 12, 1862; listed as a "rebel deserter" and took the oath of allegiance February 29, 1864.

98 - Vinson, John – enlisted as a private; absent without leave after May 1, 1863.

99 - Walls, Drewry – enlisted as a private; died of fever at Petersburg, Virginia
April 7, 1863.

100 - Walls, R.C. – enlisted as a private; in hospital January 1863 with debility.

101 - Watts, J.V. – enlisted as a corporal; wounded at Second Manassas; absent without leave after June 1, 1863.

102 - Watts, N.V. – enlisted as a private; died at Lynchburg, Virginia June 17, 1864.

103 - Watts, W. – enlisted as a private; died of chronic diarrhea in Lynchburg, Virginia June 21, 1864.

104 - Whitaker, H.H. – enlisted as a private; died of disease January 17, 1863.

105 - Whitaker, John – enlisted as a private; absent without leave after July 20, 1862.

106 - Wilkerson, E.T. – enlisted as a private; wounded and captured at Gettysburg; sent to Point Lookout, Maryland and then to Fort Delaware as a prisoner of war in July 1863.

107 - Wilkins, Robert – enlisted as a private; suffered from fever in Farmville, Virginia hospital from August 15 to September 4, 1862.

108 - Williams, Robert – enlisted as a private; absent sick during August 1862; died of disease October 29, 1862 and was buried in the Confederate Burial Grounds in Winchester, Virginia, in grave #789.

109 - York, J.A. – enlisted as a private; sick most of 1863; fought at Chickamauga; treated for a thigh wound in Danville, Virginia May 23, 1864; absent without leave after November 1, 1864.

Company G – Raised Cherokee County, Alabama and mustered into service with the regiment on May 10, 1862 at Auburn, Alabama.

1 - Alford, Jackson – enlisted as a private; absent without leave after June 1863.

2 - Andrews, Oliver P. – enlisted as a private.

3 - Arnold, James F. – enlisted as a private and promoted to First

Lieutenant; withdrawn in May 1862.

4 - Atwood, Lemuel G. – enlisted as a private and promoted to sergeant; wounded in the leg at Deep Bottom.

5 - Bellew, James M.F. – enlisted as a private; detailed as a nurse to help the wounded at Gettysburg; killed at The Wilderness May 6, 1864.

6 - Barksdale, J.A. – enlisted as a private.

7 - Batson, Henry – enlisted as a private; died of disease June 18, 1863.

8 - Battles, Lorenzo – enlisted as a private; wounded in the right hand at Second Manassas; wounded again at Gettysburg; was given special duty as an ambulance driver; paroled at Appomattox Court House.

9 - Bearden, James M. – enlisted as a private; absent without leave at the hospital in Richmond May 1, 1863.

10 - Bearden, John – enlisted as a private; absent sick without leave in May 1864.

11 - Bishop, Daniel – enlisted as a private; absent without leave after May 1863.

12 - Blackburn, Joseph – enlisted as a private.

13 - Blankenship, R.J. – enlisted as a private.

14 - Boyd, Samuel Turrentine – enlisted as a private; wounded severely in the face at The Wilderness.

15 - Brown, David S. – enlisted as a private; paroled at Appomattox Court House.

16 - Burgess, Robert – enlisted as a private; worked the entire war as a wagoneer.

17 - Burke, Francis M. – enlisted as a private at formation of company and received commission as First Lieutenant; paroled at Appomattox Court House.

18 - Campbell, Samuel D. – enlisted as a private; was ranked as "wagon master".

19 - Campbell, Thomas – enlisted as a private; killed at Cedar Run, Virginia.

20 - Cannon, Robert C. – enlisted as a private; killed at Sharpsburg.

21 - Case, L. – enlisted as a private.

22 - Chambers, Harvey R. – enlisted as a private and was promoted to sergeant; wounded in the knee at Gettysburg and taken to David's Island in New York City as a prisoner of war; died from his wounds August 19, 1863 and is buried in the Camp Hill Cemetery in grave #729.

23 - Choate, Henry J. – enlisted as a private; died of typhoid fever October 29, 1862.

24 - Coates, John R. – enlisted as a private; killed in action at Gettysburg.

25 - Conshorn, James M. – enlisted as a private; captured at Mossy Creek near Knoxville, Tennessee January 22, 1864 and sent to the Rock Island Barracks in Chicago, Illinois.

26 - Cox, W. – enlisted as a private; complained of wounds and admitted to a Meridian, Mississippi hospital January 15, 1865.

27 - Crump, John W. – enlisted as a private; killed in action at Cedar Run.

28 - Cushing, James M. – enlisted as a private; captured at Mossy Creek near Knoxville, Tennessee January 22, 1864 and sent to the Rock Island Barracks in Chicago, Illinois.

29 - Darrow, John – enlisted as a sergeant and demoted to private; wounded at Gettysburg; absent without leave after March 1864.

30 - Davis, Andrew J. – enlisted as a private; worked as a blacksmith; paroled at Appomattox Court House.

31 - Davis, Jackson D. – enlisted as a private.

32 - Fanchier, Calvin M. – enlisted as a private; discharged on January 9, 1863 suffering from rheumatism.

33 - Faughender, John C. – enlisted as a private; wounded at Cedar Run; detailed to work at division wagon train unit in July 1864; paroled at Appomattox Court House.

34 - Fletcher, S.J. – enlisted as a private.

35 - Foster, Hiram L. – enlisted as a private and promoted to corporal; sent to work at a Confederate States shoe factory in Columbus, Georgia; after spending time in a hospital there he was discharged after being declared crippled by the surgeon.

36 - Gay, Richard N. – enlisted as a private; wounded in the head at The Wilderness; paroled at Appomattox Court House.

37 - Gilbert, A.M. – selected as First Sergeant at formation of company; discharged due to disease.

38 - Gilbert, Harrison D. – enlisted as a private; died of disease at Petersburg, Virginia August 14, 1863.

39 - Gilbert, Issac – enlisted as a private and promoted to First

Joshua Price

Sergeant.

40 - Gilbert, J. – enlisted as a private; died of wounds in Knoxville, Tennessee November 29, 1863.

41 - Gilbert, Mackness W. – enlisted as a private; killed at Second Manassas.

42 - Gilbert, P.B. – elected as Captain at formation of the company May 27, 1862; died of pneumonia on May 27, 1863.

43 - Gilbert, Pinckney J. – enlisted as a private; died of disease May 28, 1863.

44 - Gilbert, Samuel N. – enlisted as a private; died of disease October 13, 1862.

45 - Gregory, Edward P. – enlisted as a private.

46 - Gregory, Griffin G. – enlisted as a private; wounded in the side October 13, 1864; worked as a blacksmith for the remainder of the war.

47 - Hale, James E. – enlisted as a private; wounded in the thigh at Gettysburg; died as a prisoner of war at Fort Delaware July 21, 1863.

48 - Hall, Andrew Jackson – enlisted as a private; worked as a regimental butcher; paroled at Appomattox Court House.

49 - Hall, J.B. – enlisted as a private; appointed company Chaplain June 24, 1862; wounded at Second Manassas; resigned from his post October 7, 1862.

50 - Hammett, Tillman – enlisted as a private; suffered from diarrhea as a prisoner of war.

51 - Hammond, J.G. – appointed First Sergeant at formation of company.

52 - Hammond, W.C. – enlisted as a private; appointed Quarter Master Sergeant.

53 - Hampton, William – enlisted as a private.

54 - Harrison, B.H. – enlisted as a private.

55 - Harriss, J.C. – enlisted as a private.

56 - Hayden, Silas E. – enlisted as a sergeant and demoted to private; present throughout most of the war; paroled at Appomattox Court House.

57 - Heald, Joseph E. – enlisted as a private; wounded at Gettysburg; died from his wounds on July 20, 1863.

58 - Helms, Charles – enlisted as a private; wounded in the big toe at The Wilderness; paroled at Talladega, Alabama May 18, 1865.

59 - Helms, Frances Marion – enlisted as a private; wounded at Suffolk, Virginia in May 1863; took oath of allegiance April 12, 1865.

60 - Henderson, William – enlisted as a private; sick most of war.

61 - Hilton, John – enlisted as a private and promoted to First Sergeant; was only seventeen years old when he enlisted in Gadsden in May 1862; was wounded at Gettysburg and paroled at Appomattox Court House.

62 - Hold, Joseph D. – enlisted as a private; killed at Chickamauga.

63 - Hood, Christopher A. – enlisted as a private; discharged on October 7, 1862 suffering from "Phthisis Pulmonalis" at a hospital in Lynchburg, Virginia.

64 - Hughes, Albert F. – elected Third Lieutenant at formation of

Joshua Price

regiment and promoted to Second Lieutenant; resigned from his post November 28, 1862.

65 - Hughes, Joseph A. – enlisted as a private; discharged.

66 - Jarvis, Fran – enlisted as a private; killed at Sharpsburg.

67 - Jarvis, J.D. – enlisted as a private; sick most of 1864.

68 - Jarvis, John J. – enlisted as a private and promoted to corporal; paroled at Appomattox Court House.

69 - Jarvis, William L. – enlisted as a corporal; killed at Sharpsburg.

70 - Johnson, Benjamin Franklin – enlisted as a private; was a courier on General Law's staff; wounded May 24, 1864 and sent home to recover; was admitted to a hospital in Richmond, Virginia suffering from gangrene.

71 - Johnson, Charles R. – enlisted as a private and promoted to sergeant; wounded at Port Royal August 25, 1863; absent without leave after The Wilderness.

72 - Johnson, F.M. – enlisted as a private; wounded and captured at Gettysburg; arrived at Fort Delaware October 15, 1863; died there November 8, 1863.

73 - Jones, Abner B. – enlisted as a private; died of complications of diarrhea in Richmond March 8, 1863.

74 - Jordan, J.J. – enlisted as a private; killed at Second Manassas.

75 - Keith, Levi – enlisted as a private; had his finger shot off May 24, 1864; paroled at Appomattox Court House.

76 - Keating, James – enlisted as a private; was assigned to the "Pioneer Corps" during May and June 1863; deserted September 27, 1864; paroled soon after that date.

77 - Kennedy, James – enlisted as a private; wounded at Second Manassas; sick in Culpepper, Virginia hospital October 20, 1862; deserted January 23, 1864; on March 21, 1865 he was captured by Union forces and sent to Louisville, Kentucky as a "rebel deserter" – where he took the oath of allegiance.

78 - Kennedy, J.R. – enlisted as a private.

79 - Kiker, Benjamin – enlisted as a private; wounded at Second Manassas; paroled at Appomattox Court House.

80 - Kingsbury, Charles E. – enlisted as a private; appointed to the Quarter Master Department; detailed to work in a hospital in Richmond.

81 - Lankford, Silas – enlisted as a private; discharged because of disease on November 1, 1862.

82 - Latham, Andrew F. – enlisted as a private; died of typhoid fever July 26, 1862.

83 - Law, James A. – enlisted as a private; wounded in the hand at Sharpsburg; paroled as a prisoner of war at Fort McHenry, Maryland.

84 - Maize, James M. – enlisted as a private; in hospital suffering of rheumatism from July 11, 1862 to July 23, 1862; killed at Second Manassas.

85 - Martin, William – enlisted as a private; died of measles June 29, 1862.

86 - Mathias, Thomas L. – enlisted as a private; worked as a teamster; died of fever February 28, 1863.

87 - May, Robert – enlisted as a private; died of disease at Gordonsville, Virginia August 26, 1862.

88 - McDaniel, George W. – enlisted as a private; worked as a teamster; killed at Deep Bottom on August 16, 1864.

89 - McDuffie, Norman E. – elected First Lieutenant at formation of company; wounded at Cedar Run; promoted to Captain October 1, 1862; wounded in the right leg at The Wilderness.

90 - McDuffie, William W. – enlisted as a private; wounded in the foot and captured at Gettysburg; sent to David's Island, N.Y.C. as a prisoner of war; recovered and was exchanged; detailed to work at division wagon train after his return to the regiment.

91 - McGill, John – enlisted as a private and promoted to corporal; discharged in February 1863 suffering from "Phthisis Pulmonalis".

92 - Means, Pleasant Barnes – enlisted as a private; wounded seriously in the right leg at Gettysburg and left on the field; his leg was amputated by a Union surgeon; he recovered from his wound and was sent home.

93 - Morange, John S. – was elected the company's first Captain; wounded severely at Cedar Run; resigned August 16, 1862.

94 - Morris, John A. – enlisted as a private.

95 - Norton, Silas B. – enlisted as a sergeant; discharged from disease September 25, 1862.

96 - Orr, J.C. – enlisted as a private; sick after 1863.

97 - Owens, Henry – enlisted as a private; suffered from diarrhea during May 1863; died June 30, 1863 in Richmond.

98 - Patterson, David M. – enlisted as a corporal and demoted to private; deserted on January 23, 1864.

99 - Patterson, Henry – enlisted as a private; died of pneumonia June 16, 1863 in Gordonsville, Virginia at the General Receiving Hospital

(also known as "Charity Hospital").

100 - Poindexter, P.P. – enlisted as a private; wounded at Chickamauga.

101 - Polk, William – enlisted as a corporal; died of disease August 30, 1862.

102 - Quishan, J.M. – enlisted as a private.

103 - Ramsey, Robert M. – enlisted as a private; paroled at Appomattox Court House.

104 - Reed, George – enlisted as a private; died of fever July 24, 1862.

105 - Reed, George H. – enlisted as a private; died of chronic diarrhea August 14, 1862.

106 - Reed, Gilmer H. – enlisted as a private; died of disease August 11, 1862.

107 - Reid, Glen H. – enlisted as a private; died of chronic diarrhea August 14, 1862.

108 - Rice, John H. – enlisted as a private; present through most of the major battles; paroled at Appomattox Court House.

109 - Rink, John T. – enlisted as a private; absent sick after June 24, 1863.

110 - Robertson, William J. – enlisted as a private; died of disease July 5, 1862.

111 - Roden, M.E. – enlisted as a private.

112 - Roden, W.M. – enlisted as a private.

113 - Rowan, Green Lewis – enlisted as a private; sick in hospital most of 1863 with chronic diarrhea; present at Chickamauga; wounded in the leg August 16, 1864; sent home to recover.

114 - Ross, Jacob – elected Second Lieutenant at formation of company; kept his rank at company's consolidation with company E July 17, 1862.

115 - Roden, J.M. – enlisted as a sergeant.

116 - Rucks, William J. – enlisted as a private; died of acute bronchitis December 25, 1862.

117 - Sands, F.M. – enlisted as a private.

118 - Sands, John – enlisted as a private.

119 - Sartin, W. – enlisted as a private.

120 - Sauls, S.E.F. – enlisted as a private; died from meningitis August 26, 1862.

121 - Shahan, Alonzo – enlisted as a private.

122 - Shirley, J.S. – enlisted as a private; died of syphilis August 4, 1862.

123 - Sibert, William J. – enlisted as a private; wounded at Second Manassas; detached as a wagon master; absent without leave after June 1863.

124 - Sitz, Andrew J. – enlisted as a private and promoted to sergeant; wounded at Gettysburg; wounded at Chickamauga; sent home to recover.

125 - Smith, L. – enlisted as a private.

126 - Smith, Otterson – enlisted as a private; appointed Second Lieutenant November 28, 1862; resigned January 1, 1864.

127 - Sotherland, E.H. – enlisted as a corporal and demoted to private; absent without leave after August 10, 1863.

128 - Sotherland, R.J. – enlisted as a private; absent without leave after October 1863.

129 - Stone, W.F. – enlisted as a private.

130 - Sutton, James T. – enlisted as a sergeant; wounded at Cedar Run; promoted to First Lieutenant October 1, 1862; wounded in the arm at Gettysburg; recovered in hospital; dropped from the roll November 15, 1864.

131 - Turley, John A. – enlisted as a private; wounded during the spring 1863; furloughed home; paroled at Appomattox Court House.

132 - Thurkill, James A. – enlisted as a private; died of disease October 18, 1862.

133 - Trotter, W.M. – enlisted as a private; worked as a teamster.

134 - Turrentine, Daniel C. – enlisted as a private; appointed Quarter Master.

135 - Turrentine, James L. – enlisted as a private.

136 - Walker, S.P. – enlisted as a private; killed at Second Manassas August 29, 1862.

137 - Weaver, J.W. – enlisted as a private; died of pneumonia as a prisoner of war January 12, 1865; prison and grave location are not listed.

138 - Wesson, James W. – appointed First Sergeant at formation of company; died of fever at West Sulfur Springs, Virginia October 25, 1862.

139 - Wilson, A.W. – enlisted as a private.

140 - Wilson, G. – enlisted as a private; was sick in a Richmond hospital July 15, 1863.

141 - Wilson, John – enlisted as a private and promoted to sergeant; wounded at Second Manassas; paroled at Appomattox Court House.

142 - Wilson, William B. – enlisted as a private; detailed to work as a hospital aide in the summer 1863.

143 - Woodliff, Augustine L. – elected First Lieutenant at formation of the company; promoted Captain August 6, 1862; resigned October 1, 1862.

144 - Wyatt, Rayburn – enlisted as a private; absent without leave during July and August 1863 "for want of proper information"; wounded at Chickamauga.

145 - Wikely, H.K. – elected Second Lieutenant at formation of company; his commission was terminated at the consolidation of companies E and G.

146 - Yates, Samuel – enlisted as a private; died at Gordonsville, Virginia August 27, 1862.

147 - Yearly, William – enlisted as a private.

Company H – Raised in Cherokee County, Alabama. Mustered into service April 29, 1862 – likely at Jacksonville, Alabama.

1 - Anderson, M.C. – enlisted as a private and promoted to sergeant; wounded at Second Manassas and at The Wilderness; paroled at Appomattox Court House.

2 - Angle, John H. – enlisted as a private.

3 - Angle, R.M. – enlisted as a private; wounded at Second Manassas.

4 - Arthur, John S. – enlisted as a private; killed by a shell August 14, 1864.

5 - Arthur, W.H. – enlisted as a private and promoted to corporal; wounded at Second Manassas; wounded severely in the shoulder on October 7, 1864 and was given furlough for the remainder of the war.

6 - Berry, Alfred Walter – enlisted as a private; taken prisoner at while fighting around Knoxville, Tennessee; is found on a list of rebel deserters who took the oath to the U.S. Government at Knoxville on December 16, 1863.

7 - Blackwell, B.F. – enlisted as a private; died of disease while on leave to Cherokee County, Alabama June 22, 1862.

8 - Blackwell, George L. – enlisted as a private; complained of "chills and fever" in October 1862; died of disease at Lynchburg, Virginia December 1, 1862.

9 - Blanton, Solomon H. – enlisted as Fourth Sergeant and promoted to First Sergeant; wounded at Second Manassas; paroled at Appomattox Court House.

10 - Brum, J.J. – enlisted as a private.

11 - Burgess, J.J. – enlisted as a private.

12 - Butler, W.H. – enlisted as a private; wounded at Second Manassas; wounded and captured at Sharpsburg; exchanged; in February 1864 he worked as a carpenter in Richmond, claimed to have been sent by General Lee to work as a guard and a nurse at a hospital in Richmond; was sent to Selma, Alabama in December

Joshua Price

1864 to work as "man in charge of government stables"; this man has a medical record that makes for one of the longest files in the compilation records; was paroled as a prisoner of war in Mobile, Alabama on May, 21, 1865.

13 - Chandler, F.M. – enlisted as a private; killed at Second Manassas.

14 - Clifton, Frances M. – elected as Third Lieutenant at formation and promoted to Second Lieutenant August 8, 1863; spent July 1864 sick in hospital.

15 - Clifton, G.T. – enlisted as a private; suffered frequently from diarrhea.

16 - Crews, John C. – enlisted as sergeant and demoted to private; detached to temporary service to the Eastern Tennessee-Virginia Railroad prior to the transfer of Longstreet's Corps south; killed at Chickamauga.

17 - Crews, Jeffrey E. – enlisted as a private and promoted to corporal; missing and declared a prisoner of war December 31, 1863; captured by Union forces in Cherokee County, Alabama on August 18, 1864 and sent to Louisville, Kentucky as a prisoner of war where he was paroled on June 17, 1865.

18 - Crews, John – enlisted as a private; wounded at Chickamauga and died from his wounds October 5, 1863.

19 - Cromer, G.A. – enlisted as a private; wounded at Second Manassas; wounded in the leg severely at Deep Bottom on August 16, 1864.

20 - Cromer, J.C. – enlisted as Third Sergeant; wounded in the knee at Cedar Run; detached service to work as a shoemaker in a Confederate shoe factory in Richmond from July to December 1864; suffered from spasms in his swollen knee in November 1864.

21 - Cromer, Thomas J. – enlisted as a private; worked as a teamster;

captured near Mossy Creek on January 22, 1864 and was sent to the Rock Island Barracks in Chicago, Illinois in February 1864; served remainder of war as a prisoner.

22 - Cromer, William F. – enlisted as a private; discharged from disease in October 1862.
Cunningham, J.E. – enlisted as a private; deserted early in the war.

23 - Day, Alfred – enlisted as a corporal and promoted to sergeant; wounded at Second Manassas; complained of rheumatism in May 1863; paroled at Appomattox Court House.

24 - Day, William – enlisted as a private; paroled at Appomattox Court House.

25 - Edwards, B.F. – enlisted as a private and promoted to corporal; spent last two years of the war sick in a hospital.

26 – Ewing, J.M. – enlisted as a private.

27 - Ewing, J.T. – enlisted as a private; killed at Cedar Run.

28 - Farmer, John – enlisted as a private; sick at the hospital from May to July 1863; deserted from the hospital July 24, 1863; no record of him ever returning to the regiment.

29 - Flannigan, W.M. – enlisted as a private; wounded at Second Manassas; suffered from rheumatism in April 1863; transferred to the Nineteenth Alabama Infantry Regiment March 24, 1864.

30 - Fortenberry, Henry – enlisted as a private; wounded at Cedar Run; wounded at Fredericksburg; deserted in May 1863; returned and released by special orders on June 27, 1863; paroled by Union Army at Lynchburg, Virginia on April 14, 1865; in this file, it is noted to "see personal papers of B.J. Law, Captain, Sixty-ninth Regiment, North Carolina Militia, 01 JUN 1863".

Joshua Price

31 - Galimore, J. – enlisted as a private; captured at Frederick, Maryland October 2, 1862 and taken to Fort Delaware, N.Y.C. as a prisoner of war.

32 - Golightly, R.C. – elected as Captain at formation of the company; killed at Sharpsburg.

33 - Golightly, Thomas – enlisted as a private; killed at Sharpsburg.

34 - Golightly, W.P. – elected as Second Lieutenant at formation of company and promoted to Assistant Quartermaster.

35 - Greenway, James H. – enlisted as a private; wounded at Suffolk in May 1863; furloughed home to recover; captured at Gaylesville, Alabama while returning to the regiment.

36 - Greenway, John H. – enlisted as a private; wounded severely on June 2, 1864; sent home to furlough; captured at Gaylesville, Alabama while returning to the regiment.

37 - Griffin, Lewis – enlisted as a corporal and promoted to sergeant; paroled at Appomattox Court House.

38 - Grogan, Albert – enlisted as a private; wounded at Second Manassas; died of disease in Staunton, Virginia on November 2, 1862.

39 - Grogan, T.B. – enlisted as a private; wounded August 14, 1864.

40 - Grogan, Wilson – enlisted as a corporal; captured near Knoxville, Tennessee December 5, 1863; sent to Camp Chase, Ohio as a prisoner of war.

41 - Hall, Lorenzo D. – enlisted as a private; present at Gettysburg; sick most of 1864.

42 - Hallett, James A. – enlisted as a private; appointed drum major; present in 1863; sick at hospital in 1864.

43 - Hardwick, E.S. – enlisted as a private and promoted to Commissary Sergeant; sick most of 1865.

44 - Hardwick, Joseph B. – elected First Lieutenant at formation of company; wounded at Cedar Run and again at Sharpsburg; absent without leave after November 1864.

45 - Hardwick, R.S. – enlisted as a private; retired September 8, 1862 due to sickness.

46 - Hardwick, William – appointed sergeant at formation of company; wounded at Sharpsburg; died October 3, 1862.

47 - Hardwick, William Mack – elected First Lieutenant at formation of company and promoted to Lieutenant Colonel October 25, 1864; captured in June 1864 and sent to Louisville, Kentucky as a prisoner of war; released after taking the oath of allegiance July 25, 1865.

48 - Harout, W.M. – enlisted as a private; wounded and captured at Sharpsburg.

49 - Harris, G.B. – enlisted as a private; taken prisoner at Spotsylvania Court House; exchanged February 13, 1865.

50 - Harris, H.W. – enlisted as a private; paroled at Appomattox Court House.

51 - Harris, J.H. – enlisted as a private and appointed musician; wounded at Cedar Run; paroled at Appomattox Court House.

52 - Harris, T.A. – enlisted as a private; absent sick 1863-64.

53 - Hart, John J. – enlisted as a private; discharged from service due to advanced age and kidney disease October 17, 1862.

54 - Harton, William M. – enlisted as a private; wounded at Second Manassas; in and out of hospitals 1863-65; paroled in Richmond,

Virginia May 2, 1865.

55 - Hendricks, B.A. – enlisted as a private; died of disease in Richmond December 9, 1862.

56 - Herring, W.J. – enlisted as a private.

57 - Herring, J.W. – enlisted as a private.

58 - Hordines, W.M. – enlisted as a private; sick at Chimborazo Hospital in Richmond on October 24, 1862.

59 - Howard, A.J. – enlisted as a private; wounded in the leg June 3, 1864; upper third of the leg was amputated; sent home to Cherokee County, Alabama in good health.

60 - Howard, Andrew – enlisted as a private; died of rheumatism October 12, 1862.

61 - Howard, J.M. – enlisted as a private; killed while on picket duty August 18, 1864.

62 - Howard, John – enlisted as a private; suffered from rheumatism and transferred to Lynchburg, Virginia hospital May 3, 1864.

63 - Howard, John W. – enlisted as a private; died of typhoid fever in Danville, Virginia October 11, 1862.

64 - Howard, J.S. – enlisted as a private; died of typhoid fever at Mount Jackson, Virginia December 17, 1862.

65 - Howard, R.R. – enlisted as a private; worked as a teamster; paroled at Appomattox Court House.

66 - Howard, Y.M. – enlisted as a private; absent without leave to the hospital in August 1864.

67 - Hurley, Robert – enlisted as a private; absent during many main battles; paroled at Appomattox Court House.

68 - Johnson, John L. – enlisted as a private.

69 - Kellett, James L. – enlisted as a sergeant at formation of the company; died February 3, 1863.

70 - Langston, James H. – enlisted as a private; wounded at Cedar Run; admitted to the hospital with smallpox in October 1862; captured at The Wilderness and sent to Point Lookout, Maryland; took oath at Elmira, New York June 14, 1865.

71 - Loftin, Askew – transferred to the company as a private; was originally a member of John P. Ralls' artillery battery; died of typhoid fever in Richmond September 12, 1862.

72 - Lumpkin, Thomas Jefferson – elected First Lieutenant at formation of the company; promoted to Captain during the Battle of Sharpsburg and was wounded at that battle; paroled at Appomattox Court House.

73 - Mathews, M.F. – enlisted as a sergeant and demoted to private; absent without leave after May 1, 1864.

74 - McBroom, Enoch V. – enlisted as a private; discharged as disabled on September 30, 1864.

75 - McBroom, J.N. – enlisted as a private; deserted in April 1863.

76 - McBroom, J.J. – enlisted as a private; sick in hospital with typhoid fever during October 1862; wounded at Fredericksburg and was absent through most of 1863 recovering; absent without leave after May 1, 1864 and was said to have been "disabled in the head".

77 - McDaniel, Z.F. – enlisted as a private; discharged on September 30, 1864.

78 - Miller, Henry M. – enlisted as a private; captured at The Wilderness; died of pneumonia in Richmond February 3, 1865.

79 - Parker, S.C. – enlisted as a private; detached service "pioneer corps"; paroled at Appomattox Course House.

80 - Pricket, William – enlisted as a private; killed at Chickamauga on September 20, 1863.

81 - Reed, J.M. – enlisted as a private; wounded seriously in the leg at Second Manassas; left leg amputated below the knee; he was discharged after recovery.

82 - Robbins, Samuel W. – enlisted as a private.

83 - Robbins, W.F. – appointed Sergeant Major at formation of company; captured at The Wilderness.

84 - Roe, John K. – enlisted as a corporal; wounded in the forearm at Sharpsburg; discharged due to disability from the wound May 25, 1863.

85 - Roe, Solomon – enlisted as a private; suffered from rheumatism during September 1862 and furloughed home to recover.

86 - Roe, William – enlisted as a private; absent sick from August to October 1863.

87 - Scott, William – enlisted as a private; died of typhoid fever in Charlottesville, Virginia July 25, 1862.

88 - Smith, David – enlisted as a private; sent to Richmond to work in a Confederate shoe factory; absent without leave after October 23, 1864.

89 - Smith, Joel B. – enlisted as a private; killed Cold Harbor on June 4, 1864.

90 - Smith, W.W. – enlisted as a private; absent without leave June 1863.

91 - Snead, G.H. – enlisted as a private; died of typhoid fever in Richmond December 20, 1862.

92 - Snead, James E. – enlisted as a private; wounded and captured at Sharpsburg; released; captured at The Wilderness.

93 - Snead, Morgan – enlisted as a private; killed at Cedar Run.

94 - Tidwell, J.J. – enlisted as a private.

95 - Thompson, J.S. – enlisted as a private.

96 - Trammell, J.J. – enlisted as a private at the age of 47; suffered from severe rheumatism in January 1863.

97 - Waters, D. – enlisted as a private; died in Richmond June 30, 1864.

98 - Watts, John H. – enlisted as a private; discharged September 26, 1862 because of his inability to keep up with the regiment.

99 - Webb, A.J. – enlisted as a private; killed at Deep Bottom.

100 - Wells, B.H. – enlisted as a private; wounded at Fredericksburg; killed accidentally at Frederick Hall, Virginia May 12, 1863.

101 - Wells, G.W. – enlisted as a private; discharged because of disease December 26, 1862; died February 28, 1863 at Cedar Bluff, Alabama.

102 - West, James M. – enlisted as a private; captured at The Wilderness; sent to Point Lookout, Maryland as a prisoner of war; died in captivity.

103 - Wilder, Jeff – enlisted as a private; disabled and transferred from the Thirtieth Georgia Infantry in September 1864.

104 - Yeargan, M.A. – appointed First Sergeant at formation of company; killed on May 18, 1864.

105 - Yeargan, W.W. – enlisted as a private; wounded at Cedar Run; listed as "sick in Yankeedom"; died November 30, 1863 suffering from chronic diarrhea.

Company I – Raised in Cherokee and Cleburne Counties in Alabama. Mustered into service April 26, 1862 – likely in Jacksonville, Alabama.

1 - Albright, N.J. – enlisted as a private.

2 - Albright, W.M. – enlisted as a private.

3 - Barnett, William R. – enlisted as a private; died of disease in 1864.

4 - Bentley, Jonathan E. – enlisted as a private; wounded at Second Manassas and Chickamauga; suffered from disease in left groin and sent to the hospital in December 1864.

5 - Black, J.R. – enlisted as a corporal and promoted to sergeant; absent without leave in July and August 1864; paroled at Appomattox Court House.

6 - Black, T.J. – enlisted as a private; killed at Chickamauga.

7 - Bocum, Josiah – enlisted as a private; captured at Sharpsburg and spent the remainder of the war as prisoner.

8 - Bowman, A.J. – enlisted as a corporal and reduced to private; captured at Richmond, Virginia April 3, 1865; surrendered as a prisoner of war at Newport News, Virginia June 25, 1865.

9 - Bowman, J.K.P. – enlisted as a private; wounded at Manassas and Chickamauga; spent a lot of time absent without leave.

10 - Braggs, M. – enlisted as a private; deserted on September 16, 1863; paroled at Talladega, Alabama June 20, 1865.

11 - Brown, Henry J. – enlisted as a private; captured while fighting near Danridge, Tennessee January 24, 1864; sent to Rock Island Barracks, Illinois February 18, 1864; exchanged; died of chronic diarrhea in a Richmond hospital May 11, 1865.

12 - Brown, S.J. – enlisted as a private.

13 - Burgess, F.M. – enlisted as a private; absent without leave after January 1863; paroled as a prisoner of war at Talladega, Alabama on May 24, 1865.

14 - Campbell, J.H. – no rank listed.

15 - Chandler, Allen W. – enlisted as a private; captured at Deep Bottom, Virginia August 12, 1864; exchanged in March 1865; captured again and surrendered April 15, 1865.

16 - Clark, J.D. – enlisted as a private.

17 - Crain, J. – enlisted as a private; died in Lynchburg, Virginia of disease November 27, 1862.

18 - Clary, J. – enlisted as a private.

19 - Dandy, W.K. – enlisted as a corporal; killed at Cold Harbor.

20 - Daverson, W.C. – enlisted as a private; died of disease at Chimborazo Hospital in Richmond, Virginia August 8, 1862.

21 - Dobbs, M.K. – enlisted as a private; deserted near Suffolk, Virginia May 4, 1863.

22 - Dobbs, Thomas – enlisted as a private; absent without leave June 11, 1863; returned to the regiment in September 1863; captured while fighting near Knoxville, Tennessee December 4, 1863 and sent to the Rock Island Barracks near Chicago, Illinois as a prisoner of war.

23 - Dolloffe, Nathan B. – enlisted as a private; died in a Richmond, Virginia hospital June 28, 1862.

24 - Dowdy, W.R. – enlisted as a private; killed at Cold Harbor.

25 - Duncan, James D. – enlisted as a private; died of disease and is buried near Richmond February 9, 1863.

26 - Edmonson, D.E. – enlisted as a private; absent without leave after March 1864.

27 - Edwards, N.C. – enlisted as a private; paroled at Talladega, Alabama June 20, 1865.

28 - Evans, Jasper – enlisted as a private; captured near Knoxville, Tennessee December 5, 1863; died April 26, 1865 at Camp Morton in Indianapolis, Indiana suffering from "inflammation of the lungs" and is buried in Greenlawn Cemetery in grave #1550.

29 - Farley, J.G. – enlisted as a private; paroled at Appomattox Court House.

30 - Ferrell, J.H. – enlisted as a private; furloughed to the hospital in August 1864.

31 - Gabrell, H.E. – enlisted as a private; wounded at Second Manassas; wounded severely in the leg at Deep Bottom.

32 - Gabrell, J. – enlisted as a private; absent without leave to the

hospital October 22, 1862.

33 - Gabrell, J.Y. – enlisted as a private; wounded at Chickamauga; absent without leave after March 1864.

34 - Garner, M.V.P. – enlisted as a private; died from typhoid fever August 11, 1862.

35 - Garrell, William – enlisted as a private; wounded at Sharpsburg; killed in action at Chickamauga.

36 - Glover, Richard T. – enlisted as a private; died of typhoid fever in the hospital October 24, 1862.

37 - Groover, A.J. – enlisted as a private; paroled at Appomattox Court House.

38 - Groover, B.F. – enlisted as a private.

39 - Groover, P.H. – enlisted as a private and promoted to corporal; wounded October 20, 1863.

40 - Groover, W.K. – enlisted as a private; promoted to Third Sergeant on August 1, 1863; wounded in the back September 3, 1864.

41 - Guff, William – enlisted as a private.

42 - Gullion, W.F. – enlisted as a sergeant; wounded at Second Manassas; paroled at Appomattox Court House.

43 - Harper, A.J. – enlisted as a private; absent without leave after June 23, 1863; paroled at Talladega, Alabama June 20, 1865.

44 - Harper, Marion – enlisted as a private; wounded at Sharpsburg; died in Richmond of smallpox on March 13, 1863.

45 - Harris, Jabez – enlisted as a private; appointed musician; wounded at Cedar Run; paroled at Appomattox Court House.

46 - Henderson, A. – enlisted as a private; paroled at Appomattox Court House.

47 - Henson, T.J. – enlisted as a private; sent home to Calhoun County, Alabama where he died of typhoid fever January 3, 1863.

48 - Hooper, O.H. – enlisted as a sergeant and demoted to private; present from 1862-63; paroled at Talladega, Alabama on May 24, 1865.

49 - Hunley, W.P. – enlisted as a private; killed at Chickamauga.

50 - Jarrett, William – enlisted as a private; sick in hospital from September to October 1862; killed at Chickamauga.

51 - Johnson, E.E. – enlisted as a corporal; promoted to Fifth Sergeant on May 10, 1863; deserted on September 16, 1863; returned and killed at Chickamauga.

52 - Johnson, J. – enlisted as a private; killed at Chickamauga.

53 - Kilgore, Jonathan W. – enlisted as a private; died of pneumonia in Lynchburg, Virginia December 26, 1862.

54 - Johnson, Warren W. – enlisted as a private; suffered from pneumonia in November 1862; paroled at Talladega, Alabama May 22, 1865.

55 - Lewis, Joseph – enlisted as a private; died of complications of diarrhea August 15, 1862.

56 - Lokey, B.R.B. – enlisted as a private.

57 - Lokey, F.M. – enlisted as a private; died of disease in Danridge, Virginia January 16, 1864.

58 - Lokey, M.C. – enlisted as a private; wounded at Sharpsburg; captured near Knoxville, Tennessee December 5, 1863; taken as a prisoner of war to the Rock Island Barracks near Chicago, Illinois; took the oath of allegiance there and was discharged to go home December 31, 1863.

59 - Lokey, N.R. – enlisted as a private; suffered from chronic diarrhea in a Petersburg, Virginia hospital in August 1863.

60 - Lusk, J.M. – elected Second Lieutenant at formation of company.

61 - Martin, G.W. – enlisted as a private; captured in Kentucky October 12, 1863; sent to Rock Island Barracks December 5, 1863; "wishes to be loyal".

62 - McGuffey, W.G. – enlisted as a private; absent after June 1864; paroled at Talladega, Alabama May 24, 1865.

63 - Mitchell, Henry D. – enlisted as a private; died November 19, 1862 at a hospital in Danville, Virginia from fever.

64 - Mitchell, J.H – enlisted as a private and promoted to sergeant; present through most the war; paroled at Appomattox Court House.

65 - Montgomery, J. – enlisted as a private.

66 - Norton, A.J. – enlisted as a private; wounded at Chickamauga; sick most of 1864; detached to light duty at a hospital September 15, 1864.

67 - Norton, M.R. – enlisted as a private; died of typhoid fever in a Richmond hospital May 24, 1863.

68 - Norwood, H. – enlisted as a private; died of disease October 15, 1864.

69 - Nunnalley, A.P. – enlisted as a First Sergeant; killed in action at Chickamauga.

70 - Nunnalley, T.J. – enlisted as a corporal and demoted to private; wounded at Second Manassas and Chickamauga; absent without leave after March 1, 1864; paroled at Appomattox Court House.

71 - Pollard, J.J. – enlisted as a private; wounded at Sharpsburg; wounded at Chickamauga; promoted to Ensign June 14, 1864; paroled at Talladega May 30, 1865.

72 - Pounds, A.P. – enlisted as a private; wounded in the ankle at Gettysburg.

73 - Pounds, Richard T. – elected Second Lieutenant at formation of company; resigned August 30, 1862 suffering from measles and a lung infection.

74 - Pounds, W.L. – elected Second Lieutenant at formation of company; resigned September 16, 1862.

75 - Pruitt, William M. – enlisted as a private; wounded at Second Manassas; paroled at Talladega, Alabama May 24, 1865.

76 - Pruett, Joshua – enlisted as a private; paroled at Talladega, Alabama May 24, 1865.

77 - Reid, Solomon – enlisted as a private; wounded at Second Manassas; sick during 1864; paroled at Appomattox Court House.

78 - Roberts, A.J. – enlisted as a private; wounded at Second Manassas and his leg was amputated; died of wounds September 11, 1862.

79 - Scott, J.H. – enlisted as a sergeant and demoted to private; wounded to Second Manassas; absent without leave after June 29, 1863.

80 - Scott, J.L. – enlisted as a private; paroled at Talladega, Alabama

May 29, 1865.

81 - Scott, William L. – elected Second Lieutenant at formation of company; resigned October 6, 1862 due to bad health.

82 - Scott, W.F. – enlisted as a private; died of typhoid fever July 25, 1862.

83 - Sizemore, James – enlisted as a private; paroled at Appomattox Court House.

84 - Story, W.B. – enlisted as a private; paroled at Appomattox Court House.

85 - Turner, J.H. – enlisted as a corporal in company E; transferred and promoted to corporal; paroled at Appomattox Court House.

86 - Vines, B.M. – enlisted as a private; absent sick without leave after March 1, 1864.

87 - Walker, L.F. – enlisted as a private; killed at Second Manassas.

88 - Wallace, T.J. – enlisted as a private; deserted May 4, 1863; court-martialed for desertion in Richmond during the fall of 1864; paroled at Talladega, Alabama June 8, 1865.

89 - Webb, John – enlisted as a private; absent without leave after May 1, 1863; paroled at Talladega May 25, 1865.

90 - Weldon, A.J. – enlisted as a corporal.

91 - Wiggonton, Green H. – enlisted as a private; killed at Gettysburg.

92 - Wiggonton, John W. – elected Captain at formation of the company; promoted to Major; commanded regiment after William Oates wounded.

93 - Wiggonton, M.Y. – enlisted as a private; wounded at Sharpsburg; promoted to First Sergeant on September 20, 1863; captured near Knoxville, Tennessee on December 4, 1863; sent to Rock Island Barracks; paroled at Talladega May 23, 1865.

94 - Williams, J. – enlisted as a private; in Richmond hospital suffering from pneumonia March 28, 1863.

95 - Williamson, Benjamin – enlisted as a private; wounded in the thigh and captured at Gettysburg; sent to David's Island, N.Y.C. as a prisoner of war; exchanged; wounded severely August 4, 1864.

96 - Williamson, J.F. – enlisted as a private; captured near Hanover Junction, Virginia May 22, 1864.

97 - Williamson, Z. – enlisted as a private; sick at hospital upon arriving in Georgia in September 1863; deserted on September 16, 1863.

98 - Wooten, H.N. – enlisted as a private.

99 - Yarbright, W.B. – enlisted as a private; paroled at Talladega on May 24, 1865.

Company K – Raised in Calhoun County, Alabama. Mustered into service in May 1862.

1 - Aderhold, Abram – enlisted as a private.

2 - Aderhold, Andrew – enlisted as a private; paroled at Appomattox Court House.

3 - Aderhold, D.H. – enlisted as a private.

4 - Aderhold, J.R. – private; promoted to sergeant; wounded at Cedar Run.

5 - Alderhalt, J.G. – enlisted as a private; killed at Cedar Run.

6 - Alverson, H.T. – enlisted as a private; killed at Gettysburg.

7 - Anderson, Benjamin F. – enlisted as a private; paroled at Appomattox Court House.

8 - Anderson, E.M. – enlisted as a private; wounded at The Wilderness.

9 - Anderson, J.C. – enlisted as a private; absent without leave after June 1862.

10 - Anderson, J.F.C. – enlisted as a private; wounded in the arm and left to the enemy at Gettysburg; arm amputated at the shoulder.

11 - Anderson, John K. – enlisted as a private; spent most of the time sick in the hospital; was wounded at Gettysburg; sent many letters home that are now an excellent source in observing disease during the Civil War; he died at home in Calhoun County, Alabama in November 1863.

12 - Anderson, Marion A. – enlisted as a private; killed at Cedar Run.

13 - Anderson, W.O. – enlisted as a private; wounded at Chickamauga; paroled at Appomattox Court House.

14 - Bartee, Thomas J. – enlisted as private.

15 - Bearden, B. – enlisted as a sergeant at formation; wounded at Cedar Run.

16 - Beaty, Jesse – enlisted as a private; killed in a railway accident on September 14, 1863.

17 - Beaty, W.A. – elected Second Lieutenant at formation of the

company; wounded at Cedar Run and Gettysburg; killed in action at Deep Bottom.

18 - Beaty, William – enlisted as a private; absent without leave at time of formation and was never paid.

19 - Blair, James – enlisted as a private; paroled at Appomattox Court House.

20 - Bradley, F.A. – enlisted as a private; wounded in the leg at The Wilderness; paroled at Appomattox Court House.

21 - Bradley, R.K. – enlisted as a private; promoted to corporal on June 16, 1863; wounded at Cedar Run; paroled at Appomattox Court House.

22 - Brown, J.B. – enlisted as a private; worked as a wagon teamster; wounded at Second Manassas.

23 - Bryant, M.L. – enlisted as a private; wounded at Chickamauga; captured near Knoxville in January 1864 and sent to Rock Island Barracks, Illinois in February 1864; died of pneumonia March 9, 1864 as a prisoner of war; is buried south of the barracks in grave #769.

24 - Buchannan, James M. – enlisted as a private; killed in action at Gettysburg.

25 - Buchannan, Jon – enlisted as a private; died of pneumonia in November 1862.

26 - Bush, James M. – enlisted as a private; wounded at Cedar Run.

27 - Crocker, S.J. – enlisted as a private; paroled at Appomattox Court House.

28 - Davis, C.C. – enlisted as a private; absent during the Battle of Gettysburg; wounded at Deep Bottom.

29 - Dickinson, A.S. – enlisted as a private; severely wounded in the hand on October 7, 1864.

30 - Dickinson, S.J. – enlisted as a sergeant; wounded at Sharpsburg; killed at The Wilderness.

31 - Duncan, J. – enlisted as a private; died of typhoid fever August 13, 1862.

32 - Edwards, William J. – enlisted as a private; wounded at Sharpsburg; absent without leave after June 1863.

33 - Elders, William – enlisted as a private; promoted to corporal on August 1, 1864.

34 - Epps, L.E. – enlisted as a private; promoted to corporal August 1, 1864; paroled at Appomattox Court House.

35 - Erwin, J.C. – enlisted as a private; captured near Knoxville December 5, 1863; sent to the Rock Island Barracks near Chicago, Illinois.

36 - Findley, John – enlisted as a private; died August 13, 1864.

37 - Findley, J.B. – enlisted as a private; "taken off the cart dead after the Battle of Cedar Run".

38 - Fisher, Rufus A. – enlisted as a private.

39 - Foster, R.A. – enlisted as a private.

40 - Fowler, Chesley B. – enlisted as a private; spent a lot of time absent without leave at hospital; paroled at Appomattox Court House.

41 - Goolsby, Isaiah – enlisted as a private; deserted March 18, 1863; captured near Chattanooga October 22, 1863; sent to

Louisville, Kentucky where he took the oath on November 1, 1863.

42 - Griffin, A.J. – enlisted as a sergeant; wounded at Cedar Run.

43 - Griffin, William – enlisted as a private.

44 - Hamilton, H.J. – enlisted as a private; absent without leave September 1863; captured and paroled at Demopolis, Alabama June 15, 1865.

45 - Hollingsworth, John – enlisted as a private; suffered from "Phthisis Pulmonalis" and discharged from service August 11, 1862.

46 - Holmes, H.T. – enlisted as a private; died in Bristol, Tennessee April 13, 1863.

47 - Holmes, John – enlisted as a private; captured at Port Royal, Virginia September 3, 1863; sent to Point Lookout, Maryland as a prisoner of war December 20, 1863; died in Washington City on December 30, 1863; was from Washington County, Alabama.

48 - Hopkins, S.T. – enlisted as a private; wounded in the hand at The Wilderness; died from complications of bronchitis in Richmond, Virginia June 25, 1864.

49 - Hubbard, John K. – elected as Captain at formation of the company; wounded in the back May 26, 1864; captured at New Market, Virginia August 15, 1864; sent to Fort Delaware as a prisoner of war in August 1864; officially dropped from regimental roll November 24, 1864; released from Fort Delaware June 17, 1865.

50 - Hubbard, J.K. – enlisted as a private; promoted to First Sergeant August 1, 1864; paroled at Appomattox Court House.

51 - Kelley, Thomas – enlisted as a private.

52 - King, D.D. – enlisted as a private; wounded from a shell to the head and right shoulder at Chickamauga; sent home to Jacksonville, Alabama to recover.

53 - Lee, J.H. – enlisted as a private; died of typhoid fever August 18, 1862.

54 - Lee, Moses – very influential soldier; elected captain at formation of the company; killed at Second Manassas.

55 - Lively, J.W. – enlisted as a private; wounded in the lung severely at Gettysburg and captured; re-captured by Confederate soldiers July 14, 1863 and recovered in Richmond, Virginia; returned to the regiment and was killed at Cold Harbor.

56 - Lowe, James M. – enlisted as a private and appointed nurse; resigned July 2, 1863; captured?

57 - Lynch, R. – enlisted as a private; died of typhoid fever December 16, 1862.

58 - Mann, W.C. – enlisted as a private; wounded and captured at Gettysburg and left on the field to the mercy of the enemy.

59 - Marable, W.R. – enlisted as a private; absent without leave sick after 1863.

60 - Martin, John – enlisted as a private; present through August 1864; sent to hospital sick; paroled at Appomattox Court House.

61 - Maryan, G.W. – enlisted as a private; died of typhoid fever in Richmond August 1, 1862.

62 - Mathis, Edward, enlisted as a private; wounded in the right arm at Second Manassas; arm was amputated and he was sent home.

63 - McAdams, James B. – enlisted as a private; died November 23, 1862.

64 - McBurnett, D.W. – enlisted as a private; paroled at Appomattox

Court House.

65 - McDuffey, P.G. – enlisted as a private; wounded and captured at Gettysburg.

66 - McElrath, William H. – enlisted as a private; suffered from heart and lung disease; detailed to work at hospital.

67 - McKinney, James K. – enlisted as a private; killed at Second Manassas.

68 - McMann, P. – enlisted as a private; captured at Chattanooga, Tennessee November 24, 1863.

69 - Mead, W.M. – enlisted as a private.

70 - Mills, H. – enlisted as a private; paroled at Talladega, Alabama May 24, 1865.

71 - Mooney, Phillip – enlisted as a sergeant.

72 - Morgan, G.W. – enlisted as a private; died in a Richmond hospital of disease August 12, 1862.

73 - Morgan, R.P. – enlisted as a corporal and promoted to Third Sergeant; spent small amount of time as a prisoner of war.

74 - Morgan, T.W. – enlisted as a private; died from "Rubiola" in Richmond, Virginia August 14, 1862.

75 - Morris, A.J. – enlisted as a private.

76 - Morris, J.B. – enlisted as a private; suffered from pneumonia in the hospital in December 1862; killed at Gettysburg.

77 - Mosley, L.E. – enlisted as a corporal; promoted to Fifth Sergeant August 1, 1863; wounded in the face at Deep Bottom; paroled at Appomattox Court House.

78 - Mulner, J.D. – enlisted as a private; died of chronic dysentery September 25, 1862.

79 - Newton, Frank – enlisted as a private; killed at Second Manassas.

80 - Newton, J. – enlisted as a private; died of chronic diarrhea August 5, 1862.

81 - Noe, James P. – enlisted as a private at age of seventeen; too sick for service and discharged to go home; paroled September 26, 1862.

82 - Nolins, H.T. – enlisted as a private.

83 - Osborne, S.T. – enlisted as a private; detached to the "pioneer corps"; wounded severely in the leg at Cold Harbor; paroled at Appomattox Court House.

84 - Owens, William F. – enlisted as a private; worked as a teamster.

85 - Palmer, K.D. – enlisted as a private; killed at Cedar Run.

86 - Parker, Felix – enlisted as a private; detailed to hospital work as a nurse; listed as a deserter September 9, 1863.

87 - Parker, Thomas – enlisted as a private; sick at hospital in 1863; captured near Knoxville December 5, 1863; sent to the Louisville Military Prison in Kentucky as a prisoner of war in late December 1863; was transferred to Fort Delaware on February 29, 1864.

88 - Peek, Dennis – enlisted as a private; wounded at Chickamauga and sent home to recover from the wounds.

89 - Pettit, H.S. – elected First Lieutenant at formation of the company; promoted to Captain August 30, 1862 after the death of

Moses Lee; wounded at Petersburg in 1864; paroled at Appomattox Court House.

90- Phillips, Christopher – enlisted as a private; killed at Second Manassas.

91 - Phillips, George M. – enlisted as a corporal and demoted to private; wounded in the wrist at Cedar Run; detailed as a nurse after the Battle of Gettysburg to care for the wounded; captured and sent to Fort McHenry on July 3, 1863; paroled on February 17, 1865.

92 - Phillips, J.A. – enlisted as a private; sick at the hospital in 1863.

93 - Phillips, J.W. – enlisted as a private; paroled at Appomattox Court House.

94 - Phillips, L.E. – enlisted as a private; deserted August 16, 1862.

95 - Phillips, M.W. – enlisted as a private; wounded in the thigh and back at Gettysburg; captured; died at Gettysburg July 28, 1863.

96 - Phillips, Thomas – enlisted as a private; absent at Richmond for court martial during first half of 1863; returned to the regiment for the Chickamauga fight and was present the rest of the war; paroled at Appomattox Court House.

97 - Quinn, Anderson – enlisted as a private; wounded at Cedar Run; wounded in the arm at Cold Harbor; died from wounds at Howard's Grove hospital June 7, 1864.

98 - Ray, James W. – enlisted as a private; wounded and captured at Gettysburg; sent to David's Island as a prisoner of war.

99 - Reaves, G.S. – enlisted as a private; wounded at Cedar Run; worked as a teamster at division wagon train; paroled at Appomattox Court House.

100 - Reaves, J.W. – enlisted as a private; wounded during the siege of Suffolk.

101 - Reed, H.C. – enlisted as a private; died in a hospital from chronic diarrhea March 17, 1863.

102 - Rice, E.T. – elected Third Lieutenant at formation of the company; promoted to First Lieutenant November 14, 1864.

103 - Rich, N.A. – enlisted as a private; wounded at Gettysburg and left on the field.

104 - Richey, Charles – enlisted as a private; seriously wounded at The Wilderness; arm amputated and furloughed home to recover; paroled at Talladega May 22, 1865.

105 - Richey, W.L. – enlisted as a private; admitted to a Richmond hospital to recover from chronic rheumatism November 28, 1863 where he died on July 2, 1864.

106 - Roberts, J.W. – enlisted as a private; killed at Deep Bottom.

107 - Roberts, W.F. – enlisted as a sergeant; present through most of the war; paroled at Appomattox Court House.

108 - Robertson, A.C. – enlisted as a private; admitted to Farmville General Hospital April 14, 1863.

109 - Robertson, J.R. – enlisted as a private; wounded at Cedar Run and Gettysburg; sent home on August 13, 1863 to recover from wounds; returned to the regiment and was retired from service December 3, 1864.

110 - Robertson, P.M. – enlisted as a private; died of chronic pneumonia at Mount Jackson, Virginia November 22, 1862.

111 - Rolades, J.G. – enlisted as a private; wounded in the shoulder and captured at Gettysburg; sent to David's Island, N.Y.C. July 17, 1863 as a prisoner of war; died from wounds July 28, 1863.

Joshua Price

112 - Seiber, George – enlisted as a private; present through most of the war; paroled at Appomattox Court House.

113 - Senley, J.H. – enlisted as a private; killed at Cold Harbor.

114 - Sexton, F. – enlisted as a private; died of disease in Danville, Virginia on November 19, 1862.

115 - Shaw, R.P. – enlisted as a private; captured in Chattanooga on November 8, 1863; sent to Camp Morton near Indianapolis, Indiana as a prisoner of war; died at Camp Morton from chronic diarrhea October 11, 1864; is buried in Green Lawn Cemetery in grave #1142.

116 - Shurbutt, E.M. – enlisted as a private; wounded at Cedar Run; admitted to hospital in Richmond suffering from pneumonia December 6, 1862 where he died on April 25, 1863.

117 - Shurbutt, H.H. – enlisted as a private; admitted to the hospital October 1, 1862 suffering from diarrhea; killed at Chickamauga.

118 - Simpson, A.H. – enlisted as a private; wounded at Gettysburg; paroled at Appomattox Court House.

119 - Simpson, D.C. – enlisted as a sergeant and demoted to private; captured after having been wounded in the head and thigh at Gettysburg; sent to David's Island, N.Y.C. as a prisoner of war; exchanged; paroled at Appomattox Court House.

120 - Simpson, E.M. – enlisted as a private; wounded at Cedar Run and Gettysburg; sent home to recover from wounds; detached service to Macon, Georgia; ordered to return to Virginia; deserted November 26, 1864.

121 - Slatton, J.J. – enlisted as a private; worked as a teamster; wounded at Second Manassas; paroled at Appomattox Court House.

122 - Slough, Abraham – enlisted as a private; wounded at Cedar

Run; died in Richmond from smallpox March 7, 1863.

123 - Smith, E.S. – enlisted as a private/musician; small amount of time sick during 1862 and 1863; present through most of the war; paroled at Appomattox Court House.

124 - Smith, J.A. – enlisted as a private; died of typhoid fever in Richmond April 9, 1863.

125 - Smith, J.R. – enlisted as a corporal and demoted to private; wounded at Cedar Run; admitted to hospital in April 1863 suffering from typhoid and diarrhea; discharged from service April 27, 1863.

126 - Smith, William – enlisted as a private; killed in action near Knoxville on November 25, 1863.

127 - Smith, W.W. – enlisted as a sergeant.

128 - Strickland, W. – enlisted as a private; paroled at Talladega May 24, 1865.

129 - Starnes, William D. – enlisted as a private; severely injured after falling from a horse August 31, 1862 and was discharged by the medical examiner.

130 - Stewart, J.A. – enlisted as a private; wounded and captured at Gettysburg; exchanged; captured at The Wilderness; sent to Point Lookout, Maryland May 15, 1864; exchanged March 14, 1865.

131 - Stewart, W.J. – enlisted as a private; died in Richmond from typhoid fever August 1, 1862.

132 - Stews, M.E. – enlisted as a private; died of disease July 7, 1862.

133 - Stine, Marcus – enlisted as a private.

134 - Stone, Richard – enlisted as a private; sick in hospital during November 1862.

135 - Thomas, Robert – enlisted as a private; killed at Cedar Run.

136 - Thompson, Jackson – enlisted as a private; killed at Cedar Run.

137 - Treadwell, E.E. – enlisted as a private; wounded at Cedar Run; served with the provost guard during the fall of 1864; paroled at Appomattox Court House.

138 - Vauley, J. – enlisted as a private; died chronic diarrhea June 6, 1864.

139 - Wallace, J.M. – enlisted as a private; was a teamster; killed at Chickamauga.

140 - Walls, J.M. – enlisted as a corporal; demoted to private June 15, 1863; worked as a teamster; wounded in the leg and captured at Gettysburg; sent to David's Island, N.Y.C. as a prisoner of war; paroled at Talladega May 24, 1865.

141 - Waugh, R.K. – enlisted as a private; died of disease in Richmond March 23, 1863.

142 - Wells, John W. – enlisted as a private; sick/absent most of the war in the hospital; listed as surrendering to the Tenth Michigan Cavalry at Newton, North Carolina sometime around April 19, 1865.

143 - White, J.W. – enlisted as a private; paroled at Appomattox Court House.

144 - Woodall, W.T. – enlisted as a private; paroled at Talladega June 16, 1865.

Soldiers whose companies are unknown:

1 - Giles, W.J. – enlisted as a private; killed at Sharpsburg.

2 - Marcon, M. – enlisted as private; killed at Sharpsburg.

3 - Peswood, William – unknown rank; died before February 13, 1863.

4 - Read, R.D. – enlisted as a private.

5 - Roberts, C.R. – enlisted as a private.

6 - Smith, James R. – enlisted as a private; took the oath at Decatur, Alabama on June 25, 1865.

Regimental Officers - Headquarters – Forty-eighth Alabama

Sheffield, James Lawrence – Colonel; resigned May 31, 1864.

Oates, William Calvin – Colonel of the Fifteenth Alabama Regiment; commanded both the Forty-eighth and the Fifteenth from June 1864 until he was replaced by John W. Wiggonton in October 1864.

Hughes, Abner A. – Lieutenant Colonel – commanded the regiment in the absence of Colonel Sheffield from August 9, 1862 – September 18, 1862; resigned post October 12, 1862; succeeded by Jesse Alldredge.

Alldredge, Jesse J. – Lieutenant Colonel; wounded at Second Manassas through both thighs; resigned in June 1863.

Alldredge, Enoch – Major – wounded at Cedar Run; resigned in

Joshua Price

October 1862.

St. John, Columbus B. – Second Lieutenant to Major.

Wiggonton, John W. – Captain to Major; promoted to major June 17, 1864.

Eubanks, Thomas James – Adjutant; promoted to Captain and replaced by
Henry S. Figures May 2, 1863; killed in action at Wauhatchie Creek.

Figures, Henry Stokes – Adjutant; killed at The Wilderness on May 6, 1864; replaced by Frank N. Kitchell.

Kitchell, Frank N. – Adjutant – present for duty "in trenches near Richmond" October 30, 1864; succeeded Henry S. Figures.

Turrentine, Daniel C. – appointed Regimental Quartermaster June 24, 1862; resigned June 25, 1863; replaced by W.P. Golightley.

Golightley, W.P. – Regimental Quartermaster.

Penn, James – Surgeon; resigned January 21, 1863; replaced by J.M. Lowe.

Lowe, J.M. – Surgeon.

Doyle, John N. – Surgeon; paroled at Appomattox Court House.

Butler, M.A. – Assistant Surgeon; dropped from service Feb. 6, 1864.

Cane, R.W. – Assistant Surgeon; resigned January 21, 1863.

Cole, W.H. – Assistant Surgeon; sick in hospital 1863.

Butler, M.A. – Assistant Surgeon; dropped from service Feb. 6,1864.

Sergeant, H.H. – Assistant Surgeon

Lucy, W.E. – Chaplain – appointed February 16, 1863; died in May 1863.

Price, Blackford – Chaplain; commissioned June 1, 1863.

The Forty-eighth Alabama Infantry Regiment, C.S.A., 1862-65

APPENDIX D
SELECT BURIAL SITES

This list is not complete. These are all the burial sites located since 2002.

ADERHOLT, J. R.   Co. K   06/05/1828 to 01/16/1906
Post Oak Springs Baptist Church in Calhoun County, Alabama

ALLDREDGE, Andrew Jackson   Co. A   02/27/1829 to 09/04/1902
Salem Primitive Baptist Church in Blount County, Alabama

ALLDREDGE, Enoch   Co. A   05/16/1807 to 11/22/1879
Alldredge Family Cemetery in Blount County, Alabama

ALLDREDGE, Jesse J.   Co. A   08/16/1831 to 06/07/1870
Alldredge Family Cemetery in Blount County, Alabama

ALLDREDGE, Nathan   Co. A   1820 to 1875
Sivley Cemetery in Blount County, Alabama

ALVERSON, H. T.   Co. K   died 06/03/1864
Old City Cemetery, Confederate Section, Lynchburg, Virginia

AMOS, James Willie   Co. E   1844 to 04/25/1915
High Point Cemetery in Marshall County, Alabama

ANDERSON, John M.   Co. K   03/03/1836 to 11/11/1863
Nances Creek Methodist Church in Calhoun County, Alabama

ANDERSON, Marion Alexander   Co. K   died 08/09/1862
Memorial marker at Lebanon Cemetery in Cherokee County,

Alabama

ANDERSON, Marcus Calhoun   Co. H   01/07/1841 to 11/28/1915
Alfalfa Cemetery, Northeast Section, Row 20 in Caddo County, Oklahoma

ANDERSON, Thomas J.   Co. E   1845 to 1920
Diamond Cemetery in Marshall County, Alabama

ANGLE, John   Co. H   no known dates, died after 1907
Buried near Centre in Cherokee County, Alabama

ARTHUR, William H.   Cpl., Co. H   11/05/1844 to 12/14/1930
Eastland Cemetery in Eastland County, Texas

BAILEY, Jesse F.   Co. D   03/06/1837 to 06/08/1897
Berry Cemetery, Red Hill Baptist Church in Marshall County, Alabama

BALLARD, Louis J.   Co. F   10/10/1810 to 08/12/1889
Union Hill Methodist Church in Cullman County, Alabama

BALLARD, Sherard   Co. F   1845 to 1862
Memorial marker at the Union Hill Methodist Church in Cullman County, Alabama

BARNARD, Patton A.   Co. D   04/01/1823 to 07/12/1902
Oleander Cemetery Marshall County

BARTLETT, George W.   Co. C   1839 to 12/12/1894
Aurora Cemetery in Etowah County, Alabama

BATSON, William Henry   Co. G   died 06/18/1863
Maplewood Cemetery in Gordonsville, Virginia

BATTLES, Lorenzo   Co. G   07/20/1840 to 01/12/1916
Reeves Grove Baptist Church in St. Clair County, Alabama

BATY, William   Co. K   1846 to 1915
Cedar Hill Cemetery in Jackson County, Alabama

BEAM, Hiram Kelly   Co. A   died 02/28/1894
Beam Cemetery near Hon in Scott County, Arkansas

BEARDEN, Samuel Bennett   Co. B   1842 to 09/04/1862
University of Virginia Confederate Cemetery in Charlottesville, Virginia

BEAVERS, Elbert   Co. F   12/29/1830 to 11/06/1917
Union United Methodist Church in Rockdale County, Georgia

BENEFIELD, J. M.   Co. I   died on 09/16/1864
Hampton National Cemetery in Hampton, Virginia

BENTLEY, Richard P.   Co. I   no known dates
Lebanon Methodist Church Cemetery in Cleburne County, Alabama

BILLINGSLEY, Henry A.   Co. H   died in 1902
Buried in Cherokee County, Alabama

BLANTON, Solomon   Co. H   05/12/1842 to 02/28/1918
Isom's Chapel Cemetery in Limestone County, Alabama

BOOKER, S. M.   Co. A   died 10/21/1862
Old City Cemetery, Confederate Section, Lynchburg, Virginia

BOWMAN, Vince P.   Co. D   08/19/1862
Old City Cemetery, Confederate Section, Lynchburg, Virginia

BOYD, Samuel Turrentine   Co. G   05/27/1845 to 11/30/1911
Buried at Cloud Chief, Oklahoma

BRACKETT, Jesse   Co. B   10/11/1834 to 02/14/1916
Kyuka Cemetery in Etowah County, Alabama

BRADLEY, Frank A.   Co. K   08/10/1848 to 02/20/1922
Fairview Methodist Church in Cherokee County, Alabama

BRADLEY, Robert D.   Co. K   10/22/1844 to 01/29/1928
Seven Springs Cemetery in Calhoun County, Alabama

BROWN, Hiram   Co. I   died 05/11/1865
Hollywood Cemetery in Richmond, Virginia

BRYANT, M.L.   Co. K   died 03/09/1864
Rock Island POW Camp Cemetery, Grave 769, Rock Island, Illinois

BUCHANAN, J. C.   Co. K   died 11/07/1862
Stonewall Cemetery in Winchester, Virginia

BURGESS, Francis Marion   Co. I   04/15/1835 to 08/15/1892
Mount Ararat Cemetery in Clay County, Alabama

CARDWELL, John James Co. D   1834 to 07/02/1863
Killed at Gettysburg and buried there.

CARR, A. M. Floyd   Co. C   12/28/1822 to 07/24/1904
Morris Cemetery, Main Section, Jefferson County, Alabama

CARR, Henry Greenwood   Co. C   12/28/1837 to 02/14/1908
First Methodist Church Cemetery, Blountsville in Blount County, Alabama

CASH, Peter   Co. E   died 09/01/1863
Cypress Hill National Cemetery, Grave 833, Brooklyn, New York

CHADWICK, Stephen   Co. D   dates not known
Oak Woods Cemetery, Confederate Mound, Chicago, Illinois

CHAMBERS, Harvey R.   Co. G   died 08/02/1863
Cypress Hills National Cemetery, Grave 729, Brooklyn, New York

CHRISTOPHER, Lindsey Lawson   Co. A   01/20/1824 to

06/09/1904
Basin Springs Cemetery in Grayson County, Texas

CHUMLEY, George William "Young"   Co. B   10/12/1845 to 12/25/1943
Providence Hill Cemetery in DeKalb County, Alabama

CLIFTON, Francis Marion   Co. H   12/17/1836 to 02/04/1909
Harris Cemetery in Cherokee County, Alabama

CLIFTON, George T.   Co. H   04/26/1831 to 06/24/1897
Harris Cemetery in Cherokee County, Alabama

COFFEY, L. D.   Co. F   10/08/1824 to 10/10/1887
Montgomery Cemetery, Lawrence County, Alabama

COLLINS, Jerry A.   Co. C   03/10/1844 to 03/06/1923
Albertville City Cemetery in Marshall County, Alabama

COOK, J. J.   Co. A   died 10/10/1863
Oakland Cemetery in Atlanta, Georgia.

COOK, James B.   Co. A   06/02/1834 to 05/23/1915
Collinsville City Cemetery in DeKalb County, Alabama

COOK, Robert   Co. F   died 12/03/1863
Point Lookout Cemetery, Point Lookout, Maryland

COOPER, Jasper Newton   Co. B   1822 to 02/20/1863
Noble Hill Cem., Etowah Co., Alabama.

CORBIN, James   Co. D   dates not known
Old City Cem., Confederate Section, Lynchburg, Virginia

CORBIN, William Riley   Co. E   04/1845 to after 1910
Rice Cemetery in Marshall County, Alabama

CORNELIUS, Madison F.   Co. C   1825 to 1890
Rock Springs Baptist Church Cemetery in Marshall County, Alabama

COX, Samuel W.   Co. B   06/09/1844 to 07/09/1906
Oak Hill Cemetery in Etowah County, Alabama

CROMER, George Adam   Co. H   1824 to 08/05/1897
Buried near Cedar Bluff in Cherokee County, Alabama

CUNNINGHAM, George   Co. E   died 02/11/1865
Green Lawn Cemetery, Section 32, Grave #130, Indianapolis, Indiana

DAILY, George W.   Co. F   05/29/1845 to 06/13/1914
Austin Creek Baptist Church Cemetery in Blount County, Alabama

DAVIS, Andrew Jackson   Co. G   dates not known
Red River County, Texas

DAY, Alfred   Co. H   04/1844 to after 1910
Ami Cemetery, near Menlo in Chattooga County, Georgia

DAY, William   Co. H   died after 1898
Buried in Cherokee County, Alabama

DeARMOND, Jerome Napoleon   Co. H   killed in 1863
Wife buried in Randle Cemetery in Cherokee County, Alabama

DENSMORE, Samuel P.   Co. A   09/30/1832 to 01/29/1913
New Salem Baptist Church Cemetery in Morgan County, Alabama

DICKINSON, Lewis J.   Co. K   died 05/09/1864
Dickinson Family Cemetery in Calhoun County, Alabama

DICKSON, Alvin Oscar   Co A   05/04/1839 to 1923
Brooksville Salem Primitive Baptist Church Cemetery in Blount

County Alabama

DIXON, J.   Co. E   died 08/04/1862
Old City Cemetery, Confederate Section, Lynchburg, Virginia

DOBBS, James M.   Co. E   04/30/1846 to 04/07/1908
Mt. Carmel Cemetery in Marshall County, Alabama

EASON, Bryant C.   Co. E   10/09/1822 to 07/29/1902
Eason - Johnson Cemetery in Marshall County, Alabama

ELDERS, William   Co. K   1841 to 1925
Bethel Cemetery in Calhoun County, Alabama

ELLIS, Reuben   Co. F   09/30/1812 to 08/09/1876
First Methodist Church in Blount County, Alabama

EUBANKS, Thomas James   Co. D.   died 10/30/1863
Has memorial markers at the Guntersville City Cemetery, at a private cemetery in Warrenton, and also in Rome, Georgia. Eubanks' final resting place is in the private cemetery in Warrenton – which was near his home.

EWING, Reuben T.   Co. H   died 05/28/1902
Tate's Chapel Cemetery in Cherokee County, Alabama

EVANS, Jasper   Co. I   died 04/26/1865
Greenlawn Cemetery, grave #1550, Indianapolis, Indiana

FIELD, Charles D.   Co. D   no known dates
New Harmony Baptist Church in Cullman County, Alabama

FIGURES, Henry Stokes   Adjutant   2/1844 to 05/06/1864
Maple Hill Cemetery in Madison County, Alabama. Original burial site in Virginia near The Wilderness battlefield. His family retrieved the body after the war for re-interment.

FINLEY, J. B. Co. K   no known dates
University of Virginia Confederate Cemetery in Charlottesville, Virginia

FORTNER, Levi   Co. A   1840 to 1914
New Harmony Cemetery in Union County, Mississippi

FOWLER, Chesley B. Co. K   04/11/1831 to 03/05/1887
Gibbs Chapel Cemetery in Blount County, Alabama

GALLOWAY, J.P.   Co. B   04/06/1843 to 11/01/1907
Conn Cemetery in Etowah County, Alabama

GALLOWAY, W. H.   Co. B   05/11/1844 to 03/28/1891
Brushy Creek Cemetery, Brushy Creek in Anderson County, Texas.

GARRARD, James Moorin   Co. C   02/03/1843 to 12/31/1931
Beulah Baptist Church in Marshall County, Alabama

GARRARD, John Franklin   Co. C   10/09/1840 to 1931
Shady Grove Baptist Church in Etowah County, Alabama

GILBERT, Alexander M.   Co. G   01/23/1838 to 05/24/1908
Black Oak Cemetery in DeKalb County, Alabama

GILBERT, Phillip B.   Co. G   01/13/1822 to 02/13/1900
Black Oak Cemetery in DeKalb County, Alabama

GILBERT, Pinckney J.   Co. G   1834 to 05/28/1863
Black Oak Cemetery in DeKalb County, Alabama (memorial marker)

GILBERT, Samuel N.   Co. G   1839 to 10/13/1862
Stonewall Cemetery in Winchester, Virginia. Memorial marker in the Black Oak Cemetery in DeKalb County, Alabama.

GILBERT, W. W.   Co. E   04/08/1845 to 01/24/1922
Lusk Chapel Methodist Church Cemetery in DeKalb County,

Alabama

GLAISE, S.   Co. C   died 10/25/1862
Stonewall Cemetery in Winchester, Virginia

GRAVES, Cunningham   Co. A   08/02/1843 to 12/06/1905
Holly Pond Cemetery in Cullman County, Alabama.

GRAVES, Hasten W.   Co. A   1839 to 01/28/1898
Lotus Grove Cemetery in Madison County, Alabama

GRAVES, Houston F.   Co. A   08/05/1842 to 07/09/1929
Gibbs Chapel Cemetery in Blount County, Alabama

GREEN, James Pinkney   Co. A   10/03/1829 to 08/10/1910
Hopewell Baptist Church Cemetery in Cullman County, Alabama

GREENE, Lewis Rowan   Co. G   1829 to 1908
Copeland Bridge Cemetery in DeKalb County, Alabama

GRIFFIN, Lewis   Co. G   died circa 1870
Buried in Cherokee County, Alabama

GROOVER, Phillip Henry   Co. I   05/19/1839 to 08/19/1911
Wesley Chapel Cemetery in Cleburne County, Alabama

GROOVER, William Kirby   Co. I   03/11/1836 to 11/11/1917
Antioch Cemetery in Cleburne County, Alabama

GROSS, Jonathan W.   Co. D   07/26/1836 to 07/31/1904
Ridgeway Cemetery in Marshall County, Alabama

GUNTER, James   Co. A   1830 to 1870
Dickson Cemetery near Brooksville in Blount County, Alabama

HAMILTON, William   Co. F   died 10/16/1863
Finn's Point National Cemetery in Salem, New Jersey

HAMMER, D. W.   Co. B   died 01/02/1863
Hollywood Cemetery in Richmond, Virginia

HAMMETT, Daniel Wilson   Co. G   1830 to 01/01/1863
Hollywood Cemetery in Richmond, Virginia

HAMMETT, Perry   Co. B   04/28/1833 to 1922
Buffington Memorial Methodist Church Cemetery in Etowah County, Alabama

HAMMETT, Tillman   Co. G   07/04/1834 to 08/31/1916
Buffington Memorial Methodist Church Cemetery in Etowah County, Alabama

HAMMOND, William C.   Co. G   no dates on headstone
Evergreen United Methodist Church Cemetery in St. Clair County, Alabama

HARDWICK, Elijah S.   Co. H   1831 to 1897
Bankhead Cemetery in DeKalb County, Alabama

HARDWICK, Joseph B.   Co. H   1842 to 01/30/1866
Cedar Bluff Cemetery in Cherokee County, Alabama

HARDWICK, William McTyiere "Mack"   Co. H   02/10/1834 to 05/16/1919
Adonirum Cemetery in Henry County, Alabama

HARDWICK, William Watt   Co. H   01/05/1841 to 10/03/1862
Cedar Bluff Cemetery in Cherokee County, Alabama (memorial marker)

HARRIS, J. C.   Co. C   06/22/1829 to 10/28/1914
Pioneer Cemetery in Eastland County, Texas

HAYES, Oliver P.   Co. E   1823 to 1862
Skirum United Methodist Church Cemetery in DeKalb County, Alabama (marker)

HELMS, Francis M.   Co. G   dates not known
Mt. Tabor Cemetery in Marshall County in Alabama

HENDERSON, J. D.   Co. F   10/28/1829 to 09/18/1891
Tidwell Cemetery in Blount County, Alabama

HENDERSON, Thomas Patrick   Co. F   born 12/28/1831
Johnson Grove Cemetery in Cullman County, Alabama

HESTER, J. W.   Co. C   06/15/1820 to 11/02/1908
New Prospect Cemetery in Marshall County, Alabama

HICKS, B. F.   Co. A   04/20/1864
Emory & Henry College Cemetery in Emory, Virginia

HITT, W. N.   Co. B   1837 to 1913
Piggott Cemetery in Clay County, Arkansas

HOLMES, John   Co. K   died 12/30/1863
Point Lookout Cemetery at Point Lookout, Maryland

HOOD, Edward Columbus   Co. A   1844 to 1864
Hood Cemetery in Blount Co., Alabama (memorial marker)

HOOPER, Obadiah H.   Co. I   1844 to 1939
Fairview Cemetery at Randlett in Cotton County, Oklahoma

HORTON, J. N.   Co. B   10/18/1842 to 03/09/1889
Bethlehem Church Cemetery in Marshall County, Alabama

HORTON, S. E.   Co. C   died 11/13/1863
Oak Hill Cemetery in Coweta County, Georgia

HOWARD, J.S.   Co. H   died 12/17/1862
Our Soldiers Cemetery at Mount Jackson in Shenandoah County, Virginia

HUBBARD, F. M.   Co. D   died 08/06/1862
Old City Cemetery, Confederate Section, Lynchburg, Virginia

HUFFSTUTLER, John   Co. A, D   07/18/1826 to 09/19/1863
Tabernacle Methodist Church in Blount County Alabama

HUGHES, Joseph   Co. B   03/14/1842 to 10/14/1921
Forrest Cemetery in Etowah County, Alabama

JARRETT, J.J.   1822 to 01/09/1863
Oakland Cemetery in Atlanta, Georgia

JENKINS, Lewis M.   Co. A   died 06/12/1863
Maplewood Cemetery in Gordonsville, Virginia

JOHNSON, F. M.   Co. G   died 11/08/1863
Point Lookout Cemetery at Point Lookout, Maryland

JOHNSON, John R.   Co. C   born 1837
Happy Home Baptist Church near Guntersville in Marshall County, Alabama

JOHNSON, Warren W.   Co. I   04/24/1836 to 12/25/1893
Rice Cemetery near New Market in Madison County, Alabama

JOHNSON, William P.H.   Co. E   08/19/1842 to 11/27/1903
Fairview Cemetery in Marshall County, Alabama

JONES, Abner B.   Co. G   died 03/08/1863
Hollywood Cemetery in Richmond, Virginia

JONES, Dan M.   Co. E   no known dates
Shiloh Cemetery in Madison County, Alabama

JONES, John   Co. C   died on 07/03/1862
Hollywood Cemetery in Richmond, Virginia

JONES, William M.   Co. E   died after 1864
Hollywood Cemetery in Richmond, Virginia

KAY, Alexander   Co. B   died 08/25/1862
Thornrose Cemetery in Staunton, Virginia

KAY, John C.   Co. B   died 02/1905
Buried in Cherokee County, Alabama

KEY, Charles Wesley   Co. B   08/18/1828 to 12/10/1902
Hickory Grove Cemetery in Lamar County, Texas

KINNEBREW, H. C.   Co. C   07/19/1801 to 04/13/1896
Clear Springs Cemetery in Marshall County, Alabama

KING, D. R.   Co. B   died 11/26/1862
University of Virginia Cemetery, Charlottesville, Virginia

KING, James W.   Co. D   died in 1866
Warrenton Cemetery in Marshall County, Alabama

KNIGHT, Sampson G.   Co. E   1835 to 1863
Liberty United Baptist Church Cemetery in DeKalb County, Alabama

KUYKENDALL, Absolom   Co. E   1838 to 1862
Myrtletree Missionary Baptist Church in Marshall County, Alabama (memorial marker)

LANKFORD, Elijah   Co. B   03/03/1842 to 03/19/1924
Fairview Baptist Church Cemetery in Etowah County, Alabama

LANKFORD, Silas M.   Co. B   03/21/1842 to 09/02/1902
Double Springs Cemetery in Oktibbeha County, Mississippi

LASELY, William H.   Co. E   died 09/01/1863

Joshua Price

Cypress Hill National Cemetery, New York

LATHAM, Andrew F.   Co. G   died 07/26/1862
University of Virginia Cemetery in Charlottesville, Virginia

LATHAM, Elias H.   Co. A   03/15/1845 to 12/09/1911
Salem Primitive Baptist Church Cemetery in Blount County, Alabama

LAW, James A.   Co. G   01/20/1847 to 04/24/1917
Forrest Home Cemetery in Marshall County, Alabama

LAW, Robert C.   Co. E   11/09/1840 to 01/01/1925
Old Union Cemetery at Grant in Marshall County, Alabama

LAWING, Jethro   Co. C   died in 1862 or 1863
Old City Cemetery, Confederate Section, Lynchburg, Virginia

LAWSON, John   02/08/1928
Jones Cemetery in Jackson County, Alabama

LEE, J. H.   Co. K   died 08/18/1862
University of Virginia Cemetery in Charlottesville, Virginia

LeFOY, Demarcus   Co. B   06/1829 to 1905-1910
MaCauley's Chapel Cemetery in Etowah, County Alabama

LINDSEY, Lawson C.   Co. A   01/20/1824 to 06/09/1904
Basin Springs Cemetery in Grayson County, Texas

LUCY, William Edward   Chaplain   1824 to 1863
Buried in Virginia

LUSK, M.   Co. I   died 10/19/1863
Marietta Confederate Cemetery in Cobb County, Georgia

MARTIN, F. P.   Co. B   no known dates
Providence Baptist Church Site Cemetery #1, Anniston Army Depot in Calhoun County, Alabama

MARTIN, H. C.   Co. D   died 12/22/1863
Stonewall Cemetery, Row 3, Grave 19, LaGrange, Georgia

MARTIN, William M.   Co. C   no known dates
Rock Springs Baptist Church Cemetery in Marshall County, Alabama

MATTHEWS, Levi P.   Co. F   died 01/22/1864
Nashville City Cemetery, Grave 6157 in Nashville, Tennessee

MAYFIELD, H. M.   Co. B   died 06/10/1864
Oakwood Cemetery in Montgomery, Alabama

McANALLY, John C.   Co. B   03/29/1836 to 03/25/1920
Phil Campbell City Cemetery in Franklin County, Alabama

McDUFFIE, Norman Henry   Co. G   1828 to 11/02/1885
Forrest Cemetery at Gadsden in Etowah County, Alabama

McDUFFIE, William Wallace   Co. G   10/18/1839 to 08/07/1897
Pilgrim's Rest Cemetery in Etowah County, Alabama

McELRATH, William H.   Co. K   12/10/1839 to 08/19/1916
Fairview Methodist Church Cemetery in Cherokee County, Alabama

MEAD, T. A.   Co. A   died 10/23/1862
Huguenot Springs Confederate Cemetery, Huguenot Springs, Powhatan County, Virginia

MEANS, Pleasant Barnes   Co. G   1844 - 1886
Fairview Baptist Church Cemetery in Etowah County, Alabama

MILLER, Eldredge W.   Co. C   07/14/1839 to 09/27/1903
Union Grove #2 Missionary Baptist Church in Blount County, Alabama

MILLER, Henry M.   Co. H   died 02/03/1865
Woodlawn Confederate Cemetery in Elmira, New York

MILLER, Jesse F.   Co. C   01/11/1829 to 06/13/1903
Beulah Baptist Church Cemetery in, Marshall County, Alabama

MILLER, Needham E.   Co. C   1830 to 1907
Zion Hill Primitive Baptist Church Cemetery in, Marshall Co., Alabama.

MITCHELL, John   Co. A   05/16/1840 to 08/15/1897
New Hope Cemetery #2 in Cullman County, Alabama

MITCHELL, Jonathan H.   Co. I   1811 to 1891
Buried in Brazos County, Texas

MITCHELL, Seaborn J.   Co. D   1835 to 1920
Shiloh Baptist Church Cemetery in Etowah County, Alabama

MOONEY, Phillip   Co. K   1824 to 02/18/1905
Rock Mills Cemetery in Randolph County, Alabama

MORGAN, Alfred W.   Co. C   1840 to 10/31/1863
Marietta Confederate Cemetery in Cobb County, Georgia. Confederate memorial marker at the Rock Springs Baptist Church Cemetery in Marshall County, Alabama.

MORGAN, Reuben Phillip   Co. K   10/26/1829 to 11/08/1892
Nances' Creek Methodist Church Cemetery in Calhoun County, Alabama

MORGAN, Squire   Co. B   died in 1911
Mt. Pisgah Cemetery in Etowah County, Alabama

MULNER, J. D.   Co. K   died 09/25/1862
Old City Cemetery, Confederate Section, Lynchburg, Virginia

NATION, David   Co. A   02/06/1838 to 08/29/1897
Blue Springs (New Lebanon) Cemetery in Blount County, Alabama

NORWOOD, J. C.
Old City Cemetery, Confederate Section, Lynchburg, Virginia

NUNNELLEY, Thomas L.   Co. I   05/05/1837 to 03/21/1904
Pilgrims Rest Cemetery in Cullman County, Alabama

ORR, Jonathan Gibson   Co. G   06/17/1834 to 12/11/1905
Orr Cemetery in Morgan County, Alabama

OWEN, William B.   Co. I   06/10/1845 to 10/14/1907
Holland Family Cemetery in Blount County, Alabama

OWENS, Benjamin Franklin   Co. I   09/04/1840 to 03/29/1923
Heflin City Cemetery in Cleburne County, Alabama

OWENS, W. T.   Co. K   06/04/1837 to 07/17/1922
Cane Creek Cemetery in Calhoun County, Alabama

PARRISH, Isaac M.   Co. C   1836 to 08/09/1862
Memorial marker at Rock Springs Baptist Church in Marshall County, Alabama

PARRISH, James Madison   Co. C   1837 to 07/25/1863
Cypress Hills National Cemetery, Grave 684, Brooklyn, New York

PARRISH, Thomas Jefferson   Co. C   09/11/1830 to 09/12/1915
Pleasant Site Cemetery in Franklin County, Alabama

PARRISH, William A. T.   Co. C   03/15/1828 to 08/30/1862
Memorial marker at Rock Springs Baptist Church in Marshall County, Alabama

PARTIN, George W.   Co. A   02/01/1863

Maplewood Cemetery, Gordonsville, Virginia

PENDERGRASS, P. H.   Co. E   died 08/11/1862
Old City Cemetery, Confederate Section, Lynchburg, Virginia

PHILLIPS, James Wesley   Co. K   02/24/1844 to 09/30/1929
Hebron Baptist Church Cemetery in Calhoun County, Alabama

POLK, William   Co. G   1838 to 08/30/1862
Buried in Virginia.

POLLARD, John J.   Co. I   06/22/1826 to 03/03/1896
Lebanon Methodist Church Cemetery in Cleburne County, Alabama

POUNDS, Merriman   Co. I   09/16/1833 to 12/25/1903
Lebanon Cemetery in Cleburne County, Alabama

POUNDS, Richard Franklin   Co. I   04/19/1832 to 01/10/1887
Lebanon Cemetery in Cleburne County, Alabama

POUNDS, William Lafayette   Co. I   05/21/1835 to 06/30/1994
Lebanon Cemetery in Cleburne County, Alabama

PRUITT, William T.   Co. C   02/19/1836 to 11/20/1932
Sand Rock Cemetery in Cherokee County, Alabama

PUGH, William Green   Co. C   03/03/1832 to 08/04/1904
Bethlehem Methodist Church Cemetery in Blount County, Alabama

RADLIFF, Levi   Co. F   07/03/1863
University of Virginia Cemetery in Charlottesville, Virginia

RAGSDALE, George W.   Co. F   died 05/02/1866
Buried in Cherokee County, Alabama

RAYBURN, Samuel King   Co. B   10/15/1811 to 07/15/1892
Guntersville City Cemetery in Marshall County, Alabama

REED, George   Co. G   died 07/25/1862
Hollywood Cemetery in Richmond, Virginia

REED, Jim   Co. H   no known dates
Mobile National Cemetery in Mobile County, Alabama

REED, Joseph Montgomery   Co. H   07/14/1844 to 03/18/1924
Cedar Bluff Cemetery in Cherokee Co., Alabama

REED, Solomon   Co. I   01/01/1821 to 08/06/1905
Camp Creek Cemetery in Cleburne County, Alabama

REID, Glenn H.   Co. G   10/28/1839 to 08/14/1862
Reid Family Cemetery in Chambers County, Alabama

REID, H. C.   Co. K   died 06/01/1863
Oakwood Cemetery in Montgomery, Alabama

RHODES, William H.   Co. I   05/21/1836 to 11/17/1925
Glenn Cemetery in Hood County, Texas

RICE, D. W.   Co. D   died 11/21/1863
Oak Hill Cemetery in Coweta County, Georgia

RICE, Ervin Foster   Co. K   12/06/1827 to 03/23/1907
Goosepond Cemetery in Jackson County, Alabama

RICHARDS, Thomas   Co. I   died 09/18/1863
Camp Chase Cemetery, Row 2, No. 1, Grave 24, Columbus, Ohio

ROBERTS, A. J.   Co. I   died 09/11/1862
Manassas Cemetery, Warrenton, Virginia

ROBERTSON, P. M.   Co. K   died 11/22/1862
Our Soldiers Cemetery, Mt. Jackson, Shenandoah County, Virginia

ROBINSON, William B.   Co. E   03/16/1837 to 12/10/1907
Walker Cemetery, Marshall County, Alabama

RODEN, Bennett H.   Co. E   01/04/1836 to 08/18/1903
Roden Cemetery in Marshall County, Alabama

RODEN, Francis M.   Co. E   1839 to 1865
Liberty United Baptist Church, DeKalb County, Alabama

RODEN, John B.   Co. E   10/27/1838 to 07/04/1913
Red Apple Baptist Church in Marshall County, Alabama

RODEN, Portland   Co. E   08/02/1829 to 12/26/1902
Roden Cemetery in Marshall County, Alabama

RODEN, Lafayette   Co. B   04/02/1845 to 12/13/1896
New Hope Missionary Baptist Church in DeKalb County, Alabama

RODEN, Lee   Co. G   1836 to 1915
Waurika Cemetery, Jefferson County, Alabama

ROE, John K.   Co. H   10/25/1829 to 02/08/1894
Myrtletree Missionary Baptist Church in Marshall County, Alabama

ROGERS, Adolphus   Co. C   1821 to 05/15/1863
Rehobeth Baptist Church in Marshall County, Alabama

ROLADES, J. G.   Co. K   died 07/28/1863
Cypress Hills National Cemetery, Grave 695, Brooklyn, New York

ROSS, Jesse E   Co. B   12/24/1829 to 01/03/1892
Noble Hill Baptist Church in Etowah County, Alabama

ROUMINES, Thomas J.   Co. D   no known dates
Rooks Cemetery in Marshall County, Alabama

RUCKS, William J.   Co. G   12/25/1862
Hollywood Cemetery in Richmond, Virginia

SAMPSON, James C.   Co. C   03/10/1831 to 1912
Pleasant Grove Missionary Baptist Church in Marshall County, Alabama

SAULS, John W.   Co. B   03/29/1845 to 11/11/1912
Fairview Cemetery in Etowah County, Alabama

SCOTT, Calvin   Co. C   02/05/1840 to 05/14/1925
Old Clear Creek Primitive Baptist Church in Marshall County, Alabama

SCOTT, W. F.   Co. I   died 07/25/1862
University of Virginia Cemetery in Charlottesville, Virginia

SCRUGGS, James S.   Co. A   05/10/1844 to 02/01/1919
Rock Springs Baptist Church in Marshall County, Alabama

SCRUGGS, Robert Milton   Co. D   04/08/1846 to 02/14/1871
Pleasant Grove (Old Roberson) Cemetery in White County, Arkansas

SEIBER, George   Co. K   12/10/1823 to 09/26/1902
Landers Cemetery in Calhoun County, Alabama

SELF, Henry   Co. F   died 07/03/1862
Hollywood Cemetery in Richmond, Virginia

SHEFFIELD, James L.   Colonel   12/05/1819 to 07/02/1892
Oakwood Cemetery in Montgomery, Alabama

SHEPPARD, John Harris   Co. B   died in 1900
Olive Branch Methodist Church Cemetery in Miller County, Arkansas

SHININ, M. A.   died 07/08/1862
Hollywood Cemetery in Richmond, Virginia

SHIRLEY, James S.　　Co. B　　1833 to 1862
New Harmony Missionary Baptist Church in DeKalb County, Alabama

SHIRLEY, John Newton　　Co. B　　02/06/1833 to 09/09/1924
New Harmony Missionary Baptist Church in DeKalb County, Alabama
*Also has a marker in the McGavock Confederate Cemetery at the Carnton Mansion in Franklin, Tennessee

SIEBER, George　　Co. K　　12/10/1823 to 09/26/1902
Landers Cemetery in Calhoun County, Alabama

SIMPSON, A. H.　　Co. K　　01/13/1831 to 04/27/1915
Cane Creek Cemetery in Calhoun County, Alabama

SIMPSON, David C.　　Co. K　　06/13/1833 to 10/26/1924
Iron City Cemetery in Calhoun County, Alabama

SIMPSON, S. B.　　Co. E　　died 09/03/1862
Old City Cemetery, Confederate Section, Lynchburg, Virginia

SIMS, Andrew Jackson　　Co. F　　07/24/1828 to 12/20/1910
Austin Creek Cemetery in Blount County, Alabama

SLIGH, John T.　　Co. C　　died 05/12/1864
Confederate Cemetery, Row 3, No. 2, Section 12, Alabama Square, Spotsylvania Battlefield in Spotsylvania County, Virginia

SMALL, Isham B.　　Co. E　　1836 to 1864
Copeland Bridge Cemetery in DeKalb County, Alabama

SMALL, J.B.　　Co. E, G　　1836 to 06/24/1864
Rabbittown Baptist Church in Calhoun County, Alabama

SMELSER, Adam　　Co. E　　12/15/1843 to 04/09/1918
Center Grove - Yancey Cemetery in Titus County, Texas

SMITH, Elias S.   Co. K   06/05/1833 to 11/26/1915
Argo Presbyterian Church in St. Clair County, Alabama

SMITH, James   Co. F   died 12/08/1863
Finn's Point National Cemetery in Salem, New Jersey

SMITH, James W.   Co. E   09/08/1831 to 02/19/1915
Smith Chapel Cemetery in DeKalb County, Alabama

SMITH, Joel D.   Co. D   1842 to 1905
Pleasant Valley Cemetery in Cherokee County, Alabama

SMITH, Joshua A.   Co. K   died 04/09/1863
Confederate Cemetery in Knoxville, Tennessee

SMITH, William H.   Co. K   died in 1863
Confederate Cemetery in Knoxville, Tennessee

SNEAD, Morgan   Co. H   died 08/09/1862
Buried in Virginia; Memorial marker likely in Cherokee County, Alabama

SPARKS, Amon   Co. C   1842 to 05/15/1862
Boaz First Baptist Church in Marshall County, Alabama

SPARKS, James Thomas   Co. C   1831 to 1901
Rehobeth Baptist Church in Marshall County, Alabama

SPARKS, Joshua   Co. C   1839 to 04/05/1863
Boaz First Baptist Church in Marshall County, Alabama

SPARKS, M.   Co. C   03/09/1846 to 11/12/1931
Antioch Cemetery in Pickens County, Alabama

SPARKS, Marida   Co. C   04/11/1842 to 06/30/1930

Joshua Price

Madedonia Baptist Church #2, DeKalb County, Alabama

SPARKS, Moses   Co. C   02/22/1847 to 03/19/1937
Rocky Mount Cemetery in Marshall County, Alabama

STEWART, J. W.   Co. C   no known dates
Upper Green's Creek Cemetery in Erath County, Texas

STEWART, William G.   Co. A   died 06/20/1864
Old City Cemetery, Confederate Section, Lynchburg, Virginia

STEWART, William Milton   Co. D   04/20/1843 to 04/04/1911
Rock Springs Baptist Church Cemetery, Marshall Co., Alabama

STONE, William H.   Co. B   10/18/1839 to 09/04/1926
New Home Missionary Baptist Church in DeKalb County, Alabama

SUTTON, James R.   Co. E   1833 to 1898
Union Cemetery in Marshall County, Alabama

TAYLOR, John Dykes   Co. E   05/09/1830 to 05/09/1888
Guntersville City Cemetery in Marshall County, Alabama

THACKER, H. A.   Co. C   died 05/01/1864
University of Virginia Cemetery in Charlottesville, Virginia

THOMAS, J.   died 09/14/1863
Marietta Confederate Cemetery in Cobb County, Georgia. Killed in train accident.

THOMPSON, J. S.   Co. D   died 04/15/1864
Emory & Henry College Cemetery in Emory, Virginia

THORNSBURY, A.   Co. A   died 10/01/1862
Old City Cemetery, Confederate Section, Lynchburg, Virginia

TREADWELL, Emanuel Everheart   Co. K   12/27/1846 to 04/02/1912

Mabank Cemetery in Kaufman County, Texas

TROTTER, William   Co. B   11/06/1815 to 01/17/1891
Trotter Dooley Cemetery in Etowah County, Alabama

TURNER, Andrew J.   Co. D   died in 1886
Sweet Home Cemetery in Marshall County, Alabama

TYLER, Spencer C.   Co. E   1823 to 1870
Warrenton Cemetery in Marshall County, Alabama

WARD, Stokely Donaldson   Co. C   10/1840 to 08/09/1915
Enon Cemetery in Jefferson County, Alabama

WARD, Thomas B.   Co. C   no known dates
Valliant Cemetery in McCurtain County, Oklahoma

WATTS, Daniel Dodson   Co. C   1820 to 1901
Beulah Baptist Church in Marshall County, Alabama

WATTS, J. J.   Co. E   1840 to 1862
Memorial marker at Beulah Baptist Church in Marshall County, Alabama

WATTS, Jacob V.   Co. F   11/10/1836 to 03/04/1904
Old Houston Cemetery in Winston County, Alabama

WATTS, James Britton "Britt"   Co. C   no known dates
Buffington Cem., Sallisaw, Oklahoma

WATTS, N. V.   Co. F   died 06/17/1864
Old City Cemetery, Confederate Section in Lynchburg, Virginia

WELLS, John W.   Co. K   1832 to 1911
Yalobusha County, Mississippi

WEST, James M.   Co. H   died 03/28/1865

Woodlawn Confederate Cemetery in Elmira, New York

WHITAKER, H. H.   Co. F   died 01/17/1863
Old City Cemetery, Confederate Section, Lynchburg, Virginia

WIGGINTON, John William   Co. I   05/28/1828 to 09/20/1891
Lebanon Cemetery in Cleburne County, Alabama

WILLIAMS, Robert   Co. F   died 10/29/1862
Stonewall Confederate Cemetery, Grave 789, Winchester, Virginia

WILLIAMSON, James F.   Co. I   no known dates
Old Harmony Primitive Baptist Church in Cleburne County, Alabama.
WILLIAMSON, Z.   Co. I   10/22/1828 to 12/27/1922
Lowell Methodist Church in Carroll County, Georgia.

WILSON, J. W.   Co. B   died 10/13/1862
Old City Cemetery, Confederate Section, Lynchburg, Virginia

WINKLES, Richard W.   Co. E   1833 to 1903
New Prospect Cemetery in Marshall County, Alabama

YARBROUGH, Richard   Co. A   01/05/1840 to 04/08/1885
Merrill - Providence Cemetery in Blount County, Alabama

YATES, Samuel   Co. G   1822 to 1893
Chestnut Creek Baptist Church in Chilton County, Alabama

YEARGEN, W. W.   Co. H   died 11/30/1863
Point Lookout Cemetery, Point Lookout, Maryland

The Forty-eighth Alabama Infantry Regiment, C.S.A., 1862-65

APPENDIX E
OBITUARIES OF CAPTAIN RUBEN EWING, CO. C

Gadsden Times, Saturday, May 29, 1909.

Capt. Rube Ewing Has Passed Away

Pioneer Resident of Cherokee County

Goes to His Reward

A Resident of Cherokee For More Than

Fifty Years – Prominent In

Church and Lodges

Captain Rube Ewing, aged 85 years, one of the pioneer residents of Cherokee county, passed away at his home near Round Mountain Wednesday after an illness extending over many months. He suffered with stomach trouble.

The deceased was one of the earlier residents of Cherokee County, having been a constant resident for more than fifty years. He was a farmer by occupation and owned a large plantation. He was twice elected to the State legislature representing Cherokee County and always took a prominent part in politics, being a staunch Democrat. He was a Mason of considerable prominence in the lodge and was also a member of the Baptist Church in which he was a deacon. He founded Tate's Chapel more than twenty years ago.

Mr. Ewing is survived by one brother, William, and five daughters.

The funeral was held at 10 o'clock this morning from the late home to Tate's Chapel where services were conducted. The funeral was largely attended by relatives and friends of the deceased. Interment was made in the burying grounds near the chapel.

From *Cherokee Harmonizer*, Centre, AL. – Thursday, June 3, 1909.

Capt. R.T. Ewing Passes Away.

Capt. R.T. Ewing, aged 86 years, died at his home near Centre, at 12 o'clock midnight, 27th of May.

He had resided at the place where he died since soon after the Civil War. He was an ex-Confederate soldier, a Mason, a member of the Baptist Church, and a good and much respected citizen.

His whole life was a quiet and unassuming one, though he always manifested a proper interest in public affairs, and twice represented Cherokee in the Legislature.

He was a farmer by occupation, and before his declining years, carried on extensive farming operations, but of late years leased his possession.

He always, as long as he was physically able, took much interest in the affairs of the farmers and in farmer's organizations. He lived well and died in peace.

He was buried last Saturday morning at Tate's Chapel with Masonic Honors.

From the *Coosa River News*, Friday, June 11, 1909.

I could not print the obituary so I had to take notes from it.

Captain Ruben T. Ewing:

- Was made a Mason June 3, 1870.
- Served 3 terms as worshipful master
- Vocal about his Christianity
- Joined C.S.A. in 1861 as a Private
- Elected Captain of the 48 AL
- Highly respected man
- Friend to the orphans, widows, and helpless

Ewing's obituaries are interesting because he is the stereotype of an individual who embraced the Lost Cause and the people in his community considered him as being so.

APPENDIX F
JAMES SHEFFIELD MURDER TRIAL CLIPPINGS

The Gadsden Times - Thursday, June 26, 1890.

Killed his man.

Col. Sheffield Kills Dr. May at Warrenton.

The little town of Warrenton, in Marshall County, was thrown into a state of excitement last Friday morning by the killing of Dr. William May by Col. Jas. L. Sheffield at the latter's daughter.

Col. Sheffield is employed as a clerk in one of the departments at Montgomery, and last Thursday he received a telegram to come home at once, and it is said by those who know that his family's honor was involved and he went to protect it. He met Dr. May at the home of his (Sheffield's) daughter and an altercation ensued in which May was shot and killed.

After the shooting Sheffield left but returned on Saturday morning and surrendered and is now lodged in the county jail. He is 60 years old.

Col. Sheffield made no effort to escape, but quietly came forward and gave himself up. It was not known that he was in custody until he was placed in jail. He surrendered at the residence of Judge T.A. Street, four miles from Guntersville. Preliminary trial took place Monday before Hon. S.K. Rayburn. The state was represented by Solicitor Lusk, and the defendant by Messrs. Brown, Holliday, and Street. Dr. May was buried at the city cemetery with Masonic Honors.

Gadsden Times-News - July 10, 1890.

Sheffield Acquitted.

Guntersville, Ala., July 3 – Col. J.L. Sheffield, who a few days ago, shot and instantly killed Dr. May at Warrenton, near here, has been acquitted at the preliminary trial, which has been in progress several days, and left immediately for Montgomery. The evidence adduced tended to show that Dr. May was a near neighbor to Colonel Sheffield's daughter, who was an invalid and was demented. May was also her family physician, and, it is alleged, made indecent proposals to the lady, which resulted in the killing.

APPENDIX G
APPOMATTOX ROSTER

Field and Staff
Major John Wiggonton

R.W. Cane, Hospital Steward

E.S. Smith, Musician

Company A
Alvin O. Dickson, Captain

R. A. Yarborough, First Sergeant

J.S. Burgess, Second Sergeant

S.P. Densmore, Third Sergeant

Milton Stewart, Fourth Sergeant

A.J. VanHorn, First Corporal

C.H. Graves, Fourth Corporal

Whitfield Beam, Private

R.A. Cain, Private

F.G. Grigsby, Private

A. Hyatt, Private

E.H. Latham, Private

L.C. Mitchell, Private

John Mitchell, Private

William Neely, Private

William, Stepp, Private

Charles Graves, Private

Company B
Lieutenant George W. Chumley

W.N. Hitt, Second Sergeant

Thomas Langford, First Corporal

Thomas Roden, Second Corporal

M.E. Gilliland, Private

Thomas Howell, Private

C.W. Kay, Private

Issac Broom, Private

Silas Kay, Private

S.M. Langford, Private

H.F. Penn, Private

James Sitz, Private

John W. Sauls, Private

Ervin Sheffield, Private

Company C
H.C. Kinnebrew

George Washington Bartlett, First Sergeant

Needham E. Miller, Private

Thomas Jefferson Parrish, Second Sergeant

Jesse Freeman Miller, Private

J.W. Billingsly, Private

J.B. Watts, Private

G.D. McCracken, Private

Company D

Green Black, Third Sergeant

William A. Gilbert, Fourth Sergeant

W.F. Gullion, Fifth Sergeant

R.M. Scruggs, First Corporal

J.K. Widner, Third Corporal

Calvin Black, Private

G.W. Grass, Private

B.H. Renfro, Private

James Scruggs, Private

William Putnam, 4 CPL

Company E

J.W. Amos, Second Sergeant

W.L. Morris, Fifth Sergeant

Jerry M. Dobbs, Private

Richard Winkles, Private

Mack Rector, Private

T.P. Waller, Private

W.W. Gilbert

Company F

A.J. Edwards, First Sergeant

J.P. Patterson, Second Sergeant

W.T. Crawford, Fifth Sergeant

L.O. Miller, Private

J.C. Lambert, Private

T.A. Smith, Private

Company G

John Hilton, First Sergeant

John Wilson, Second Sergeant

J.J. Jarvis, First Corporal

Andrew Jackson Hall, Pvt.

Benjamin F. Kiker, Private

Levi Keith, Private

Joshua Price

Lorenzo Battles, Private

W.W. McDuffie, Private

J.D. Brown, Private

J.H. Rice, Private

Andrew Jackson Davis, Private

R.A. Ramsey, Private

R.M. Gay, Private

J.A. Turley, Private

Silas E. Hayden, Private

J.C. Forfender, Private

Company H

S.H. Blanton, First Sergeant

Robert Hurley, Private

M.C. Anderson, Third Sergeant

R.R. Howard, Private

Levi Griffin, Fourth Sergeant

J.S. Harris, Private

Alfred Day, Fifth Sergeant

S.C. Parker, Private

Hiram Wesley Harris, Private

William Day, Private

Company I

J.H. Mitchell, Fourth Sergeant

John G. Farley, Private

J.R. Black, Fifth Sergeant

A.J. Groover, Private

J.H. Turner, First Corporal

Solomon Reed, Private

R. Bentley, Third Corporal

W.B. Story, Private

A. Henderson, Fourth Corporal

James A. Sizemore, Private

The Forty-eighth Alabama Infantry Regiment, C.S.A., 1862-65

Company K

J.F. Hubbard, First Sergeant	S.F. Osborn, Private
W.F. Roberts, Fourth Sergeant	Thomas Phillips, Private
L.E. Mosely, Fifth Sergeant	J.W. Phillips, Private
L.E. Epps, Second Corporal	G.S. Reeves, Private
R.D. Bradley, Third Corporal	A.H. Simpson, Private
J.R. Aderholt, Private	D.J. Simpson, Private
Abram Aderholt, Private	George Sieber, Private
D.H. Aderholt, Private	J. Slatton, Private
B.F. Anderson, Private	E.E. Treadwell, Private
F.A. Bradley, Private	J.M.W. White, Private
James A. Blair, Private	W.O. Anderson, Private
S.J. Crocker, Private	Chesley B. Fowler, Pvt
D.W. McBurnett, Private	John Martin, Private

Please note: This roster was taken from John Dykes Taylor's work that was edited by W.S. Hoole in 1985. The roster seems close to complete, I made a few changes to it. However, there is likely a few names, such as officers, missing. Also, there may be some names that are misspelled.

Joshua Price

APPENDIX H
ROSTER OF COMPANY CAPTAINS

Company A

Andrew Jackson Alldredge

Jesse J. Alldredge

Randolph Graves

Company B

Thomas J. Burgess

Company C

W.S. Walker

J.M. Bedford

H.C. Kinnebrew

Company D

Samuel A. Cox

Thomas J. Eubanks

Company E

Samuel K. Rayburn

Phillip B. Gilbert

Isom B. Small

Company F

Reuben Ellis

Jeremiah Edwards

Company G

John S. Moragne

Augustine Woodliff

N.H. McDuffie

Company H

R.C. Golightley

William Mack Hardwick

T.J. Lumpkin

Company I

J.W. Wiggonton

Reuben T. Ewing

Company K

Moses Lee

Roster taken from Oates' The War Between the Union and the Confederacy combined with my own personal notes.

APPENDIX I
SELECT PHOTOGRAPHS OF SOLDIERS

Special appreciation goes to the following people for providing me with photographs of some of the soldiers of the Forty-eighth Alabama, and also giving me to produce them in this work:

Colonel James L. Sheffield – courtesy of Judge Bobby M. Junkins

Major John Wiggonton – courtesy of Wandagene Jones

Adjutant Henry Figures – online source

Captain Alvin Oscar Dickson – ADAH, Lottie Painter Hudson

Captain Randolph Graves – online source

Captain Reuben Ewing – ADAH

Captain Samuel Rayburn – courtesy of Larry Joe Smith

Captain Thomas Eubanks – courtesy of Lottie Painter Hudson

Captain Augustine Woodliff – courtesy of Sherry Clayton

Lt. George William Chumley – courtesy of Emily Porter

Lt. Jesse Ross – compliments of Sherry Clayton

Sgt John Newton Shirley – courtesy of Betty Shirley

Sgt Thomas J. Parrish – courtesy of Earnest E. Hill

Sgt John Dykes Taylor – courtesy of Judy Taylor Reed

Sgt Silas Norton – courtesy of Sherry Clayton

Pvt Samuel Bearden – courtesy of Earnest E. Hill

Pvt Jesse Brackett – courtesy of Georgia Brackett Jones

Pvt Calvin Scott – courtesy of Wayne & Gloria Gregg

Pvt Samuel Dunn – courtesy of ADAH

Pvt Samuel T. Boyd – courtesy of Donna Standerfer

Pvt John Anderson – courtesy of William E. Simpson

The Forty-eighth Alabama Infantry Regiment, C.S.A., 1862-65

Colonel James Lawrence Sheffield

Major John Wigginton

Adjutant Henry Stokes Figures

Captain Alvin Oscar Dickson, Co. A

Captain Randolph Graves, Co. A

Captain Reuben T. Ewing, Co. C

Captain Samuel King Rayburn, Co. D

Captain Thomas Eubanks, Co. D

The Forty-eighth Alabama Infantry Regiment, C.S.A., 1862-65

Captain Augustine Woodliff, Co. G

Lieutenant George Chumley, Co. B

The Forty-eighth Alabama Infantry Regiment, C.S.A., 1862-65

Lieutenant Jesse Ross, Co. B

Sgt. John Newton Shirley

Sgt. Thomas Jefferson Parrish, Co. C

Sgt. John Dykes Taylor, Co. D

The Forty-eighth Alabama Infantry Regiment, C.S.A., 1862-65

Sgt. Silas B. Norton, Co. G

Pvt. Samuel Bearden, Co. B

Pvt. Jesse Brackett, Co. B

Pvt. Calvin Scott, Co. C

Pvt. Samuel C. Dunn, Co. E

Pvt. Samuel T. Boyd, Co. G

Pvt. John M. Anderson, Co. K

BIBLIOGRAPHY

PRIMARY SOURCES AND MANUSCRIPTS

Dickson, Alvin Oscar. *Letter*. Alabama Department of Archives and History.

Figures, Henry Stokes. *Letters, 1861-1864*. Gettysburg National Military Park.

Hood, John Bell. *Advance and Retreat: Personal Experiences in the United States and Confederate States Armies.* New Orleans: Hood Orphan Memorial Fund, 1880.

Hoole, William Stanley, ed. *John Dykes Taylor's History of the 48th Alabama Volunteer Infantry Regiment, C.S.A.* Tuscaloosa: Confederate Publishing Company, 1985.

Hoole, William Stanley, ed. "The Letters of Captain Joab Goodson, 1862-1864." *The Alabama Review 10*, no.2 (April 1957): pages 126-153.

Longstreet, James. *From Manassas to Appomattox: Memoirs of the Civil War in America.* Philadelphia: J.B. Lippincott Company, 1896.

Oates, William Calvin. *Lost Opportunities: The War Between the Union and the Confederacy.* Dayton, Ohio: Press of Morningside Bookshop, 1985.

Simpson, William E., ed. *The Correspondence of John M. Anderson, Private, C.S.A., 1862-1863.* Privately published, no date. Houston Cole Library, Jacksonville State University.

Stocker, Jeffrey D., ed. *From Huntsville to Appomattox: R.T. Coles' History of 4th Regiment, Alabama Volunteer Infantry, C.S.A., Army*

of Northern Virginia.
Knoxville: University of Tennessee Press, 1996.

Turner, Charles W., ed. "Major Charles A. Davidson: Letters of a Virginia Soldier." *Civil War History 22*, no.1 (March 1976): 16-40.

U.S. War Department. *The War of the Rebellion: A Compilation of the Official Records of the Union and Confederate Armies.* One Hundred twenty-eight Volumes total, Washington, D.C., 1880-1901.

Vaughan, Turner. "The Diary of Turner Vaughan, Co. C., Fourth Alabama Regiment, C.S.A." *Alabama Historical Quarterly 18*, no. 4 (December 1956): pages 573-601.

Woodliff, Augustine. *Letters, 1862-1865*. In possession of Northeast Alabama Genealogical Society, Nichol's Library, Gadsden, Ala.

NEWSPAPERS

Confederate Veteran.

(Guntersville) *Democrat.*

Jacksonville Republican.

Philadelphia Weekly Press.

(Richmond) *Enquirer.*

GENEALOGICAL REFERENCE BOOKS

The Heritage of Blount County, Alabama. Clanton, Alabama: Heritage Publishing Consultants, Inc., 1999.

The Heritage of Marshall County, Alabama. Clanton, Alabama: Heritage PublishingConsultants, Inc. 2000.

SECONDARY SOURCES

Ambrose, Stephen E. "Yeoman Discontent in the Confederacy." *Civil War History 8*, no. 3 (September 1962): pages 259-268.

Bowles, Pinckney, "Battle of Cold Harbor", *Philadelphia Weekly Press*, January 31, 1885.

Casler, John O. *Four Years in the Stonewall Brigade.* Marietta: Continental Book Co., 1951.

Catton, Bruce. *A Stillness at Appomattox*. New York: Doubleday and Company, Inc., 1953.

Collins, George K. *Memoirs of the One Hundred and Forty-ninth Regiment New York Volunteer Infantry.* Syracuse, New York: by the author, 1891.

Cozzens, Peter. *The Shipwreck of Their Hopes: The Battles for Chattanooga.* Chicago: University of Illinois Press, 1994.

Cozzens, Peter. *This Terrible Sound: The Battle of Chickamauga.* Chicago: University of Illinois Press, 1992.

Cunningham, Horace H. *Field Medical Services at the Battles of Manassas.* Athens: University of Georgia Press, 1968.

Cunningham, Horace H. *Doctors in Gray: The Confederate Medical Service.* Baton Rouge: Louisiana State University Press, 1993.

Dorman, Lewy. *Party Politics in Alabama from 1850 to 1860.* Tuscaloosa: The University of Alabama Press, 1935.

Douglas, Henry Kyd. *I Rode with Stonewall.* Chapel Hill: University of North Carolina Press, 1945.

Frank, Joseph Allan and George A. Reaves. *"Seeing the Elephant":*

Raw Recruits at the Battle of Shiloh. New York: Greenwood Press, 1989.

Gaff, Alan D. *On Many a Bloody Field: Four Years in the Iron Brigade.* Indianapolis: Indiana University Press, 1996.

Greene, A. Wilson. "Opportunity to the South: Meade versus Jackson at Fredericksburg." Civil War History 33, no. 4 (December 1987): pages 295-314.

Hahn, Stephen. *Roots of Southern Populism: Yeoman Farmers and the Transformation of the Georgia Upcountry, 1850-1890.* New York: Oxford University Press, 1984.

Henderson, G.F.R. *Stonewall Jackson and the American Civil War.* New York: Barnes and Noble, Inc., 2006.

Hennessy, John J. *Return to Bull Run: The Campaign and Battle of Second Manassas.* New York: Simon and Schuster, 1993.

Horn, Stanley F. *The Army of Tennessee.* Norman: University of Oklahoma Press, 1952.

Huffman, James. *Ups and Downs of a Confederate Soldier.* New York: William E. Rudge's Sons, 1940.

Johnson, Timothy D. and Guy R. Swanson. "Conflict in East Tennessee: Generals Law, Jenkins, and Longstreet." *Civil War History 31,* no.2 (June 1985): 101-110.

Krick, Robert K. *Stonewall Jackson at Cedar Mountain.* Chapel Hill: The University of North Carolina Press, 1994.

Laine, J. Gary and Morris M. Penny. *Law's Alabama Brigade in the War Between the Union and the Confederacy.* Shippensburg: White Mane Publishing Co., 1996.

Laine, J. Gary and Morris M. Penny, *Struggle for the Round Tops: Law's Alabama Brigade at the Battle of Gettysburg, July 2-3, 1863.*

Shippensburg: White Mane Publishing Company, 1999.

Law, Evander M., "The Struggle for Round Top," *Battles and Leaders of the Civil War, Volume III.* New York: The Century Company, 1888.

Lord, Walter, ed., *The Fremantle Diary: Being the Journal of Lieutenant Colonel James Arthur Lyon Fremantle, Coldstream Guards, on his three months in the Southern States.* Boston: Little Brown Publishing Co., 1954.

Martin, Bessie. *A Rich Man's War, A Poor Man's Fight: Desertion of Alabama Troops from the Confederate Army.* Tuscaloosa: University of Alabama Press, 2003.

McPherson, James M. *Battle Cry of Freedom: The Civil War Era.* New York: Oxford University Press, 1988.

McPherson, James M. *Crossroads of Freedom: Antietam, The Battle that Changed the Course of the Civil War.* New York: Oxford University Press, 2002.

McPherson, James M. *For Cause and Comrades: Why Men Fought in the Civil War.* New York: Oxford University Press, 1997.

Moore, Albert Burton. *Conscription and Conflict in the Confederacy.* New York: Hillary House Publishers, Ltd., 1963.

Nolan, Alan T. *The Iron Brigade: A Military History.* New York: MacMillan, 1961.

Nosworthy, Brent. *The Bloody Crucible of Courage: Fighting Methods and Combat Experiences of the Civil War.* New York: Carroll and Graf Publishers, 2003.

Ownby, Ted. Subduing Satan: *Religion, Recreation, and Manhood in the Rural South, 1865-1920.* Chapel Hill: University of North

Carolina Press, 1990.

Pfanz, Harry. *Gettysburg: The Second Day.* Chapel Hill: The University of North Carolina Press, 1987.

Purcell, Douglas Clare. "Military Conscription in Alabama During the Civil War". *Alabama Review,* (April 1981): 94-106.

Rable, George C. *Fredericksburg! Fredericksburg!* Chapel Hill: The University of North Carolina Press, 2002.

Rhea, Gordon C. *Cold Harbor: Grant and Lee, May 26 to June 3, 1864.* Baton Rouge: Louisiana State University Press, 2002.

Rhea, Gordon C. *The Battles for Spotsylvania Court House and the Road to Yellow Tavern: May 7-12, 1864.* Baton Rouge: Louisiana State University Press, 1997.

Rhea, Gordon C. *The Battle of the Wilderness: May 5-6, 1864.* Baton Rouge: Louisiana State University Press, 1994.

Rhea, Gordon C. To *The North Anna River: Grant and Lee, May 13-25, 1864.* Baton Rouge: Louisiana State University Press, 2000.

Robertson, James I., Jr. *The Stonewall Brigade.* Baton Rouge: The Louisiana State University Press, 1963.

Sears, Stephen W. *Landscape Turned Red: The Battle of Antietam.* New York: Houghton-Mifflin Co., 1983.

Skinner, George W., ed. *Pennsylvania at Chickamauga and Chattanooga: Ceremonies at the Dedication of the Monuments Erected by the Commonwealth of Pennsylvania.* William Stanley Ray, State Printer of Pennsylvania, 1904.

Smith, Larry Joe. *Guntersville Remembered.* Albertville, Alabama: Creative Printers, Inc.: 1989.

Smith, William R. *The History and Debates of the Convention of Alabama, 1861*. Montgomery: White, Pfister, and Company, 1861.
Storey, Margaret M. "Civil War Unionists and the Political Culture of Loyalty in Alabama, 1860-1861." *Journal of Southern History 69*, no. 1 (February 2003): pages 71-106.

Wert, Jeffry D. *A Brotherhood of Valor: The Common Soldiers of the Stonewall Brigade, C.S.A., and the Iron Brigade, U.S.A.* New York: Simon and Schuster, 1999.

Wiley, Bell I. *The Life of Johnny Reb and The Life of Billy Yank*. Baton Rouge: Louisiana State University Press, 1994.

Worsham, John H. *One of Jackson's Foot Cavalry: His Experience and What He Saw During the War, 1861-1865*. New York: Neale Publishing Co., 1912.

Wyatt-Brown, Bertram. *Southern Honor: Ethics and Behavior in the Old South*. New York: Oxford University Press, 1982.

Yoseloff, Thomas. *The Official Atlas of the Civil War*. New York, 1958.

Joshua Price

Vitae

Joshua Price was born in Gadsden, Alabama in 1978 to Jimmy and Carolyn Price. He attended West End High School (Walnut Grove) where he graduated in 1996. He attended Jacksonville State University, where he earned a B.A. in History (2008) and a M.A. in History (2010).

Mr. Price is a secondary education teacher in Blount County, Alabama at Susan Moore High School. He has also served as an Adjunct Instructor of History for Snead State Community College.

He and his wife Alysia, along with their 11-year old son Jackson and four-year old daughter KaytiAnn, make their home in the Susan Moore Community.

Mr. Price is currently working on a regimental history of the Nineteenth Alabama Infantry Regiment, C.S.A., 1861-65 – of which he began research in 2011 and has collected hundreds of documents, letters, diaries, and photographs related to the experiences of the soldiers of that regiment since that time that will be used in the narrative. A publication date for this work will be announced at a later date.

More from the author:

The Civil War Letters of John M. Anderson, Company K, 48th Alabama Infantry Regiment (2011). Joshua Price, editor.

To order an autographed copy of this book please email the author at:

Joshua Price

The Forty-eighth Alabama Infantry Regiment, C.S.A., 1862-65

Joshua Price

www.ingramcontent.com/pod-product-compliance
Lightning Source LLC
Chambersburg PA
CBHW072002150426
43194CB00008B/972